D1709224

Selected Titles in This Series

12 **N. V. Krylov,** Lectures on elliptic and parabolic equations in Hölder spaces, 1996
11 **Jacques Dixmier,** Enveloping algebras, 1996 Printing
10 **Barry Simon,** Representations of finite and compact groups, 1996
9 **Dino Lorenzini,** An invitation to arithmetic geometry, 1996
8 **Winfried Just and Martin Weese,** Discovering modern set theory. I: The basics, 1996
7 **Gerald J. Janusz,** Algebraic number fields, second edition, 1996
6 **Jens Carsten Jantzen,** Lectures on quantum groups, 1996
5 **Rick Miranda,** Algebraic curves and Riemann surfaces, 1995
4 **Russell A. Gordon,** The integrals of Lebesgue, Denjoy, Perron, and Henstock, 1994
3 **William W. Adams and Philippe Loustaunau,** An introduction to Gröbner bases, 1994
2 **Jack Graver, Brigitte Servatius, and Herman Servatius,** Combinatorial rigidity, 1993
1 **Ethan Akin,** The general topology of dynamical systems, 1993

Lectures on Elliptic and Parabolic Equations in Hölder Spaces

N. V. Krylov

Graduate Studies
in Mathematics

Volume 12

1991 *Mathematics Subject Classification*. Primary 35-01;
Secondary 35J15, 35K10.

ABSTRACT. The lectures concentrate on some basic facts and ideas of the modern theory of linear elliptic and parabolic equations in Hölder spaces and lead the reader as far as possible in a short course. We show that this theory, including some issues of the theory of nonlinear equations, is based on some general and extremely powerful ideas and some *simple* computations. The Sobolev-space theory, which basically follows the same lines, is left beyond the scope of the book (apart from the sharp Sobolev embedding theorem).

The main object of studies is the first boundary-value problems for elliptic and parabolic equations, with some guidelines concerning other boundary-value problems such as the Neumann or oblique derivative problems or problems involving higher-order elliptic operators acting on the boundary. Numerical approximations are also discussed.

The presentation has been chosen in such a way that after having followed the book the reader should acquire good understanding of what kinds of results are available and what kind of technique is used to obtain them.

For graduate students and scientists in mathematics, physics, and engineering interested in the theory of partial differential equations.

Library of Congress Cataloging-in-Publication Data

Krylov, N. V. (Nikolaĭ Vladimirovich)
Lectures on elliptic and parabolic equations in Hölder spaces / N. V. Krylov.
p. cm. — (Graduate studies in mathematics, ISSN 1065-7339; v. 12)
Includes bibliographical references and index.
ISBN 0-8218-0569-X (alk. paper)
1. Differential equations, Elliptic. 2. Differential equations, Parabolic. 3. Generalized spaces.
I. Title. II. Series.
QA377.K758 1996
515′.353—dc20 96-19426
CIP

∞ The paper used in this book is acid-free and falls within the guidelines
established to ensure permanence and durability.

10 9 8 7 6 5 4 3 2 01 00 99 98 97

Contents

Preface

In 1961, when I was a student of the fourth year at the Moscow State University (approximately the level of the second–year graduate program in the USA), I attended a regular course in the theory of partial differential equations. Later in the graduate school of the Moscow State University while dealing with some problems of optimal control of random processes, I felt a need for some results in PDEs and started to read books and articles on the subject. I was just astonished by the fact that the knowledge I got before was of no help whatsoever. Nothing like this happened with probability theory, measure theory or functional analysis. I did not even have any idea as to what constituted the framework of "modern" (at that time) theory. The basic notions of *a priori estimates* and Hölder and Sobolev spaces were not included in the standard course. The basic *method of continuity* allowing one to prove the solvability of equations on the grounds of a priori estimates was not included either. Only Laplace's equation and the heat equation were considered. What happens if we add into the equations even lower–order terms was left behind the scene. All this despite the fact that the theory of elliptic and parabolic PDEs has always been one of the basic tools in mathematics, physics, engineering and other areas of applications of mathematics.

Of course, these were the times when many important aspects of the theory were just being developed and had not yet been set down in a more or less canonical form. However, even then the well–known book by Miranda on elliptic equations and the expository article by Il'in, Kalashnikov and Oleinik on parabolic ones were already available. The excellent books by Gilbarg and Trudinger [**6**], Friedman [**7**], Ladyzhenskaya and Ural'tceva [**10**], Ladyzhenskaya, Solonnikov and Ural'tceva [**11**] came somewhat later.

Thirty–four years later in 1995 I got to teach a two–quarter course on elliptic and parabolic equations at the University of Minnesota. It is hard to describe my surprise when I found out that there was still no short albeit self–contained *textbook* for graduate students which would cover the same basic notions and answer the same basic questions without going into heavy computations. On the eve of the coming millennium I did not want to teach a course which was obsolete even 34 years ago, and this forced me to type my own lecture notes for students. The present book is a very slight extension of the lecture notes of this two–quarter course (although I added quite a few new exercises).

I have to say at once that this book is not designed as an introduction to or a guidebook on the general theory of partial differential equations. My goal was not to try to cover as many subjects as possible but rather to concentrate on some basic facts and ideas of the modern theory of elliptic and parabolic equations and to lead the reader as far as possible in a short course. The presentation has been chosen in such a way that after having followed the book, the reader should acquire a good understanding of what kinds of results are available and what kind of technique is used to obtain them. In applications this is very important knowledge, since one cannot always find in the literature all the small modifications or particular cases of known results. I also hope that after following this book the reader who decides to become a professional in the field will be well prepared for reading research articles, monographs and books such as Gilbarg and Trudinger [**6**], Friedman [**7**], Ladyzhenskaya and Ural'tseva [**10**], Ladyzhenskaya, Solonnikov and Ural'tseva [**11**], Krylov [**8**] and others containing not only mathematical results but also historical remarks and extensive bibliographies. Some of them are listed in the bibliography at the end of this book.

I set myself the task of presenting basic aspects of the theory of elliptic and parabolic equations in *Hölder spaces*, including some issues of the theory of nonlinear equations. I intend and hope to show that this theory is based on some general and extremely powerful ideas and some *simple* computations. We only deal with the Hölder–space theory of the first boundary–value problems for elliptic and parabolic equations, providing only some guidelines concerning other boundary–value problems, such as the Neumann or oblique derivative problems or problems involving higher–order elliptic operators acting on the boundary.

Very important Sobolev–space theory is left beyond the scope of this book (we only show the sharp Sobolev embedding theorem). However, the reader should not have any psychological difficulty in studying the Sobolev–space theory, since it follows the same lines and the technique is sometimes even simpler. We can recommend the very interesting recent books by Giaquinta [**4**] and Caffarelli and Cabré [**3**] bearing on this subject.

It is worth noting that usually one obtains a priori estimates in the Hölder spaces after rather lengthy, although instructive, examinations of Newtonian potentials. While treating elliptic equations in domains or parabolic equations in cylinders, one used to consider also other kinds of potentials. It turns out that application of a beautiful idea of Safonov makes the use of potentials absolutely unnecessary. Another innovation of the book is that we prove the existence theorems by using a method, introduced by Browder (1959), which does not require integral representations or Hilbert–space theories and works in the same way for domains in Euclidean spaces and for *smooth manifolds*. These two ingredients make the exposition short and self–contained and allow one to obtain the main results quickly.

Sometimes people get discouraged seeing only existence and uniqueness theorems and theorems that the solution is smoother if data is smoother. That is why we also present "real" numerical methods of approximating the solutions and show that they work owing to "abstract" existence and uniqueness results. We also discuss advantages and disadvantages of explicit formulas for solutions with regard to general theorems.

The book is designed as a textbook. Therefore it does not contain any new theoretical material but rather a new compilation of some known facts, methods and ways of presenting the material. For example, a part of Chapter 2 is close to the corresponding sections in the books by John [**5**], Gilbarg and Trudinger [**6**] or in many other books. Other parts of the book come from many other sources. The author does not make even an attempt to list all of these sources, hoping that the clear statement of not having any pretension to originality is a sufficient excuse.

My experience of following these notes at the University of Minnesota and in the Summer School at Cortona (Italy) in 1995 shows that students accept the subject very well and learn the material quickly and well. Partly this is due to exercises which have been suggested as homework.

There are about 190 exercises in the book, a few of which (about 40 marked with an *) are used in the main text. These are the simplest ones. However, many other exercises are quite difficult, despite the fact that solutions are almost always short. Therefore the reader should not feel upset if he/she cannot do them even after a good deal of thinking. Probably hints for them should have been provided right after each exercise. We do give hints to the exercises but only at the end of each chapter just to give readers an opportunity to test themselves.

Several words about notation. The index is at the end of the book. In addition, we always use the summation convention and allow constants denoted by N, usually without indices, to vary from one appearance to another even in the same proof. If we write $N = N(...)$, this means that N depends only on what is inside the parentheses. Usually in the parentheses we list the objects which are fixed. In this situation one says that the constant N is *under control*. By domains we mean general open sets. On some occasions, we allow ourselves to use different symbols for the same objects, for example,

$$u_{x^i} = \frac{\partial u}{\partial x^i} = D_i u, \quad u_x = \operatorname{grad} u = \nabla u, \quad u_{xx} = (u_{x^i x^j}).$$

To the instructor

The book starts with Chapter 1 on elliptic equations of any order with constant (complex) coefficients in the whole space and the proof of their solvability by means of the Fourier transform. This is an encouraging start: after one or two lectures students know that a very large class of problems is solvable in a very easy way. Further material in Chapter 1 is used later in the treatment of solvability of elliptic equations of any order with variable coefficients in the whole space. Chapter 1 and the rest of the book are almost independent. The whole chapter can be dropped if one is not interested in equations of higher order or of second order with complex coefficients. Also one then has to drop Secs. 8.2 and 8.3 on "explicit formulas" for solutions of the Cauchy problem for parabolic equations. This will not affect the rest of the material related to parabolic equations.

Chapter 2 contains a traditional discussion of Laplace's equation. The contents of it should become known to every student studying the theory of PDEs because underlying ideas very often inspired further development of the theory even if only implicitly. However, if one is interested in getting to the solvability of elliptic equations with variable coefficients as soon as possible, one can actually just drop Secs. 2.1 through 2.5. Further confinement to second–order elliptic and parabolic equations with real coefficients will then allow one to start reading the book from Sec. 2.6, thus skipping about one sixth of the book. Moreover, one can then skip Secs. 3.5 through 3.7. The necessary changes in Chapter 4 are discussed in detail there. In particular, one might prefer to derive Schauder's a priori estimates for second–order elliptic equations with variable coefficients in the same way as is done in Sec. 8.9 for parabolic equations.

I had a very pleasant experience teaching a short course in the Summer School at Cortona, where in twelve lectures I covered all of Chapter 2, then Chapters 3 and 4 with the shortcuts described above, and then Chapters 5 and 6 excluding Secs. 5.5, 6.6 and 6.7. Also in that course I confined myself to $C^{2+\delta}$–theory and did not discuss the results on better regularity of solutions from Sec. 4.2.

One more way to design a short course covering elliptic *and* parabolic equations, however, is to start with Chapter 8, where we study parabolic equations, looking into the previous chapters for some proofs which are not always repeated for parabolic equations, and to extract almost all information related to elliptic equations, since for functions independent of t parabolic equations become elliptic ones (cf., for instance, Corollary 8.10.2 and Remark 8.12.3).

Finally, if one is interested only in proving solvability in Hölder spaces of *Laplace's equation*, one need not prove the interpolation inequalities in Chapter 3 and can proceed as in Sec. 8.7, where we treat the heat equation.

With such a variety of possibilities it is hard to describe prerequisites. To read the whole book, one needs to be familiar with the Fourier transform and somewhat with distributions. The necessary facts are listed in sections in which they are used. The shortest course still requires the knowledge of the theory of integration in Euclidean spaces.

Acknowledgments. I want to express my gratitude to my students at the University of Minnesota and in Cortona for their real enthusiasm, which encouraged me to compile my lecture notes into this book, and for solving many exercises which were thus checked, many of them corrected and supplied with better hints. Special thanks in this regard are due to H. Yoo.

I am also grateful to L.C. Evans, E.B. Fabes, C.E. Kenig, M.V. Safonov, and N.S. Trudinger, whose interest and friendly support were very important to me.

Nicolai Krylov
Minneapolis, October 1995

CHAPTER 1

Elliptic Equations with Constant Coefficients in $\mathbb{R}^d$

In this chapter we study elliptic equations with constant coefficients of any order $m \geq 1$ in $\mathbb{R}^d$. The number m will be assumed to be bigger than or equal to 2 in the whole book starting with Sec. 1.4 (see however Exercises 3.7.8 and 3.7.9).

1.1. The notion of elliptic operator

Let $\mathbb{R}^d$ be a d–dimensional Euclidean space of points $x = (x^1, ..., x^d)$. Any d-tuple $\alpha = (\alpha_1, ..., \alpha_d)$ of integers $\alpha_k = 0, 1, 2...$ is called a multi–index. For a multi–index α and $\xi = (\xi^1, ..., \xi^d) \in \mathbb{R}^d$ denote

$$D_k = \frac{\partial}{\partial x^k}, \quad |\alpha| = \alpha_1 + ... + \alpha_d, \quad D^\alpha = D_1^{\alpha_1} \cdot ... \cdot D_d^{\alpha_d}, \quad \xi^\alpha = (\xi^1)^{\alpha_1} \cdot ... \cdot (\xi^d)^{\alpha_d}.$$

DEFINITION 1.1.1. Let $m \geq 1$ be an integer and a^α be some (complex) numbers given for any multi–index α with $|\alpha| \leq m$. The operator $L = \sum_{|\alpha| \leq m} a^\alpha D^\alpha$ is called (mth order) *elliptic* if both

$$\sum_{|\alpha|=m} a^\alpha \xi^\alpha \neq 0 \quad \text{for} \quad \xi \in \mathbb{R}^d \setminus \{0\}, \qquad \sum_{|\alpha| \leq m} a^\alpha i^{|\alpha|} \xi^\alpha \neq 0 \quad \text{for} \quad \xi \in \mathbb{R}^d.$$

The polynomial $p(\xi) = \sum_{|\alpha| \leq m} a^\alpha i^{|\alpha|} \xi^\alpha$ is called *the characteristic polynomial for* L. The operator $\sum_{|\alpha|=m} a^\alpha D^\alpha$ is called *the principal part of* L.

For a function $g(x)$ we define its Fourier transform $F(g) = \tilde{g}$ by

$$F(g)(\xi) = \tilde{g}(\xi) = c_d \int_{\mathbb{R}^d} e^{-ix\cdot\xi} g(x)\, dx, \quad c_d = \frac{1}{(2\pi)^{d/2}}.$$

We need the following facts about the Fourier transform. By $C_0^\infty(\mathbb{R}^d)$ we denote the space of all infinitely differentiable functions on $\mathbb{R}^d$ with compact support.

- If $g \in C_0^\infty(\mathbb{R}^d)$, then $\tilde{g}$ is well–defined and bounded. Furthermore, for any multi–index α

$$i^{|\alpha|} \xi^\alpha \tilde{g}(\xi) = F(D^\alpha g)(\xi),$$

  which implies, in particular, that $\tilde{g} \to 0$ faster than $|\xi|^{-n}$ for any $n > 0$ as $|\xi| \to \infty$.
- $\tilde{g}$ can be defined for any function $g \in L_2(\mathbb{R}^d)$ in such a way that $\tilde{g} \in L_2(\mathbb{R}^d)$ and for any $f, g \in L_2(\mathbb{R}^d)$ Parseval's identity holds

$$\int_{\mathbb{R}^d} \bar{f} g\, dx = \int_{\mathbb{R}^d} \bar{\tilde{f}} \tilde{g}\, d\xi.$$

- For any $g \in L_2(\mathbb{R}^d)$ we have

$$g(x) = c_d \int_{\mathbb{R}^d} e^{ix\cdot\xi} \tilde{g}(\xi)\, d\xi$$

almost everywhere, where the right-hand side is understood in L_2-sense.

REMARK 1.1.2. We have $p(\xi) = e^{-ix\cdot\xi} L e^{ix\cdot\xi}$. From here it is easy to get formally that $\widetilde{Lg} = p(\xi)\tilde{g}$. Indeed, use formally integration by parts and the following operator

$$L^* := \sum_{|\alpha| \le m} a^\alpha (-1)^{|\alpha|} D^\alpha,$$

which is called the *formally adjoint* operator for L. Also use $L^* e^{-ix\cdot\xi} = p(\xi) e^{-ix\cdot\xi}$. Then

$$\begin{aligned}\widetilde{Lg}(\xi) &= c_d \int_{\mathbb{R}^d} e^{-ix\cdot\xi} Lg(x)\, dx \\ &= c_d \int_{\mathbb{R}^d} g(x) L^* e^{-ix\cdot\xi}\, dx = p(\xi) c_d \int_{\mathbb{R}^d} e^{-ix\cdot\xi} g(x)\, dx.\end{aligned}$$

EXAMPLE 1.1.3. The characteristic polynomial of Laplace's operator

$$\Delta = \frac{\partial^2}{(\partial x^1)^2} + \ldots + \frac{\partial^2}{(\partial x^d)^2}$$

is $-|\xi|^2$ which is zero for $\xi = 0$. Therefore Δ is not an elliptic operator in the sense of our definition.

We will later (see Exercise 1.3.4) give a definition of homogeneous elliptic operators which embraces Laplace's operator. It is worth noting that elliptic operators in the sense of our definition are also elliptic in a broader sense of the book by L. Bers, F. John and M. Schechter [**2**]. In this book one can also find the definitions of strongly elliptic and properly elliptic operators.

The convenience of the current definition can be seen if one tries to solve the equation

$$Lu = f$$

in the whole space. Indeed, formally we have $p(\xi)\tilde{u} = \tilde{f}$, and since $p \ne 0$, $\tilde{u} = p^{-1}\tilde{f}$, which allows us to find u by using the inverse Fourier transform.

EXAMPLE 1.1.4. The operator $1 - \Delta$ is elliptic, for its characteristic polynomial equals $1 + |\xi|^2$.

EXAMPLE 1.1.5. If by p_L we denote the characteristic polynomial of L, then $p_{L_1 L_2} = p_{L_1} p_{L_2}$ (see Remark 1.1.2). Therefore if the operators L_1, L_2 are elliptic, so is their product $L_1 L_2$. In particular, the operator $(1 - \Delta)^k$ is elliptic for any integer $k \ge 1$.

EXAMPLE 1.1.6. The operators $\Delta + b^k D_k - 1$ for any constant $b \in \mathbb{R}^d$ and the operator $d^3/(dx)^3 + 1$ for $d = 1$ are elliptic. (Remember the summation convention.)

LEMMA 1.1.7. *If L is an elliptic operator, then there exists a constant $\kappa > 0$ called the constant of ellipticity such that*

$$|\sum_{|\alpha|\le m} a^\alpha i^{|\alpha|}\xi^\alpha| \ge \kappa(1+|\xi|^m) \quad \forall \xi \in \mathbb{R}^d. \tag{1.1.1}$$

Moreover, there exists $\varepsilon > 0$ such that

$$|\sum_{|\alpha|\le m} b^\alpha i^{|\alpha|}\xi^\alpha| \ge \frac{1}{2}\kappa(1+|\xi|^m) \quad \forall \xi \in \mathbb{R}^d$$

whenever $|b^\alpha - a^\alpha| \le \varepsilon$ for any α.

Proof. Observe that the function

$$f(t,\xi) = |\sum_{|\alpha|\le m} a^\alpha i^{|\alpha|} t^{m-|\alpha|}\xi^\alpha|, \tag{1.1.2}$$

is positive–homogeneous and continuous, and $f > 0$ on the unit sphere $t^2 + |\xi|^2 = 1$ in $\mathbb{R}^{d+1}$. Therefore, $f \ge \kappa$ on the sphere for a constant $\kappa > 0$. This implies $f(t,\xi) \ge \kappa(t^2+|\xi|^2)^{m/2}$ everywhere, so that $f(1,\xi) \ge \kappa(1+|\xi|^2)^{m/2}$. Now we use that for any numbers $\gamma, t \ge 0$

$$(1+t)^\gamma \ge (1+t^\gamma)/2.$$

Indeed, if $t \ge 1$, then the right–hand side is less than $(2t^\gamma)/2 = t^\gamma$, which is less than the left–hand side. If $t \le 1$, then the right–hand side is less than 1, which again is less than the left–hand side. Thus, $f(1,\xi) \ge (1+|\xi|^m)\kappa/2$.

To prove the second assertion of the lemma, use

$$|\sum_{|\alpha|\le m} b^\alpha i^{|\alpha|}\xi^\alpha| \ge |\sum_{|\alpha|\le m} a^\alpha i^{|\alpha|}\xi^\alpha| - N\varepsilon(1+|\xi|^m),$$

where the constant N depends only on d.

EXERCISE* 1.1.8. Prove that (1.1.1) implies the ellipticity of L.

In the future we will often use the following simple property of positive–homogeneous functions.

LEMMA 1.1.9. *Let $g(\eta)$ be a positive–homogeneous function of order γ given on $\mathbb{R}^n$. If g is bounded on the unit sphere, then $|g(\eta)| \le N|\eta|^\gamma$ for $\eta \ne 0$ with a constant N independent of η. If it is continuously differentiable at any point $\eta \ne 0$, then its first–order partial derivatives are positive-homogeneous functions of order $\gamma - 1$ bounded on the unit sphere, so that $|D_j g(\eta)| \le N|\eta|^{\gamma-1}$ for $\eta \ne 0$ with a constant N independent of η. In particular, for the function $g = f^{-1}$ with f from (1.1.2) we find that for any multi–index α*

$$|D^\alpha \frac{1}{p(\xi)}| \le N\frac{1}{1+|\xi|^{m+|\alpha|}} \quad \forall \xi \in \mathbb{R}^d$$

with a constant N independent of ξ.

1.2. Solvability. Green's representation

Let L be an elliptic operator. As we mentioned above we can try to find a solution u of the equation $Lu = f$ starting with the formula $\tilde{u} = p^{-1}\tilde{f}$ and expressing u as the inverse Fourier transform of the product $p^{-1}\tilde{f}$. Since the Fourier transform of a convolution is a constant times product of the Fourier transforms of the factors, it is natural to look for the function u of the form

$$u(x) = \int_{\mathbb{R}^d} f(y)G(x-y)\,dy = \int_{\mathbb{R}^d} f(x-y)G(y)\,dy, \tag{1.2.1}$$

where

$$G(y) = \frac{1}{(2\pi)^d}\int_{\mathbb{R}^d} \frac{1}{p(\xi)} e^{iy\cdot\xi}\,d\xi. \tag{1.2.2}$$

This makes natural the proof of the following theorem in which $C_b^\infty(\mathbb{R}^d)$ stands for the space of all infinitely differentiable functions with bounded derivatives (and remember, $C_0^\infty(\mathbb{R}^d)$ is its subspace of functions with compact support).

THEOREM 1.2.1. *For any $f \in C_0^\infty(\mathbb{R}^d)$ there exists a function $u \in C_b^\infty(\mathbb{R}^d)$ such that $Lu = f$ in $\mathbb{R}^d$.*

Proof. *Case $m > d$.* Define the function $G(x)$ by formula (1.2.2). By the way, observe that the integral obviously converges and defines G as a bounded continuous function. Also $p^{-1} \in L_2(\mathbb{R}^d)$ so that $F(G)$ exists in L_2-sense, and $F(G) = c_d p^{-1}$. Next let

$$u(x) := \int_{\mathbb{R}^d} f(x-y)G(y)\,dy = f * G(x).$$

Then of course $D^\alpha u = (D^\alpha f)*G$, and therefore $u \in C_b^\infty(\mathbb{R}^d)$. Finally, by Parseval's theorem

$$Lu(x) = \int_{\mathbb{R}^d} \bar{\bar{G}}(x-y)Lf(y)\,dy = \int_{\mathbb{R}^d} \bar{F}(\bar{G}(x-\cdot))(\xi)\widetilde{Lf}(\xi)\,d\xi$$

$$= \frac{1}{(2\pi)^{d/2}}\int_{\mathbb{R}^d} e^{ix\cdot\xi}\frac{1}{p(\xi)}p(\xi)\tilde{f}(\xi)\,d\xi = c_d\int_{\mathbb{R}^d} e^{ix\cdot\xi}\tilde{f}(\xi)\,d\xi = f(x).$$

In particular, for all $x \in \mathbb{R}^d$ and $f \in C_0^\infty(\mathbb{R}^d)$,

$$L\int_{\mathbb{R}^d} f(x-y)G(y)\,dy = \int_{\mathbb{R}^d} G(x-y)Lf(y)\,dy = f(x). \tag{1.2.3}$$

Case $m \le d$. Take any integer k such that $2k + m > d$ and observe that our equation also means that $(1-\Delta)^k Lu = (1-\Delta)^k f$. In connection with this define

$$G'(x) = \frac{1}{(2\pi)^d}\int_{\mathbb{R}^d} \frac{e^{ix\cdot\xi}}{p(\xi)(1+|\xi|^2)^k}\,d\xi, \quad u(x) = \int_{\mathbb{R}^d} G'(x-y)(1-\Delta)^k f(y)\,dy.$$

Then by (1.2.3) applied to the operator $L' = (1-\Delta)^k L$ instead of L we get

$$Lu(x) = \int_{\mathbb{R}^d} G'(x-y)L(1-\Delta)^k f(y)\,dy = f(x).$$

The theorem is proved.

EXERCISE 1.2.2. Find an infinitely differentiable (complex) function f for which the middle term of (1.2.3) exists but is not equal to f.

DEFINITION 1.2.3. A generalized function (in other words, a distribution) $G(x)$ is called a *Green's function* of the equation $Lu = f$ in the whole space if $LG(x) = \delta_0(x)$ or, equivalently, if equalities (1.2.3) hold.

We discuss this definition in the following section.

We need the following facts about distributions.

- Definition of a distribution f as linear functional on $C_0^\infty(\mathbb{R}^d)$ continuous in the sense that if $\phi, \phi_n \in C_0^\infty(\mathbb{R}^d)$ have supports in the same ball and $\phi_n \to \phi$ uniformly on $\mathbb{R}^d$ together with every derivative of any order, then $(f, \phi_n) \to (f, \phi)$. The notation: $(f, \phi) = \int_{\mathbb{R}^d} f(x)\phi(x)\,dx$.
- Any locally integrable function is a distribution (by the dominated convergence theorem).
- Every distribution is infinitely differentiable in the generalized sense: $(-1)^{|\alpha|}(f, D^\alpha \phi)$ is a distribution, which by definition is called $D^\alpha f$.
- If $\phi \in C_0^\infty(\mathbb{R}^d)$ and f is a distribution, then the function $(f, \phi(x - \cdot))$ is infinitely differentiable and for any multi–index α

$$D^\alpha(f, \phi(x-\cdot)) = (D^\alpha f, \phi(x-\cdot)) = (f, (D^\alpha\phi)(x-\cdot)).$$

- If f_n is a sequence of distributions, we say that it converges to a distribution f if $(f_n, \phi) \to (f, \phi)$ for any test function $\phi \in C_0^\infty(\mathbb{R}^d)$. In this case it holds that $D^\alpha f_n \to D^\alpha f$ for any multi–index α.

1.3. Green's functions as limits of usual functions

Observe that the equality $LG(x) = \delta_0(x)$ means that for any $f \in C_0^\infty(\mathbb{R}^d)$ we have

$$\int_{\mathbb{R}^d} G(y)L^* f(y)\,dy = f(0).$$

By putting here $f(x-y)$ instead of $f(y)$, we see that the equality $LG(x) = \delta_0(x)$ implies (1.2.3).

EXERCISE* 1.3.1. Prove the converse: (1.2.3) implies the equality $LG(x) = \delta_0(x)$.

The proof of Theorem 1.2.1 implies the following fact.

COROLLARY 1.3.2. *Take any $k = 0, 1, 2, \ldots$ such that $2k + m > d$ and define*

$$G'(x) = \frac{1}{(2\pi)^d}\int_{\mathbb{R}^d} \frac{1}{p(\xi)(1+|\xi|^2)^k} e^{ix\cdot\xi}\,d\xi, \quad G(x) = (1-\Delta)^k G'(x).$$

Then G is a Green's function for L.

REMARK 1.3.3. There is a more intuitively clear formula for G which looks like $G = c_d F^{-1}(p^{-1})$. Namely, in the sense of distributions

$$G(x) = \frac{1}{(2\pi)^d} \lim_{R\to\infty} \int_{|\xi|\le R} \frac{1}{p(\xi)} e^{ix\cdot\xi}\, d\xi. \tag{1.3.1}$$

Furthermore, for any multi–index α and integer $r = 0, 1, 2, \ldots$

$$D^\alpha G(x) = \frac{1}{(2\pi)^d} \frac{1}{|x|^{2r}} \lim_{R\to\infty} \int_{\mathbb{R}^d} e^{ix\cdot\xi} (-\Delta)^r \left[\zeta_R(\xi)\, \frac{i^{|\alpha|}\xi^\alpha}{p(\xi)}\right] d\xi, \tag{1.3.2}$$

where ζ is any function of class $C_0^\infty(\mathbb{R}^d)$ such that $\zeta(0) = 1$ and $\zeta_R(\xi) = \zeta(\xi/R)$.

To prove (1.3.1) notice that by Corollary 1.3.2 (always in the sense of distributions)

$$G(x) = \frac{1}{(2\pi)^d}(1-\Delta)^k \lim_{R\to\infty} \int_{|\xi|\le R} \frac{1}{p(\xi)(1+|\xi|^2)^k} e^{ix\cdot\xi}\, d\xi$$

$$= \frac{1}{(2\pi)^d} \lim_{R\to\infty} (1-\Delta)^k \int_{|\xi|\le R} \frac{1}{p(\xi)(1+|\xi|^2)^k} e^{ix\cdot\xi}\, d\xi,$$

which gives the result, since $(1-\Delta)^k e^{ix\cdot\xi} = (1+|\xi|^2)^k e^{ix\cdot\xi}$. To prove (1.3.2) observe that

$$|x|^{2r} e^{ix\cdot\xi} = (-\Delta_\xi)^r e^{ix\cdot\xi}, \tag{1.3.3}$$

where the subscript ξ means that Δ is applied with respect to the variable ξ. Then

$$G(x) = \frac{1}{(2\pi)^d} \lim_{R\to\infty} (1-\Delta)^k \int_{\mathbb{R}^d} \zeta_R(\xi)\, \frac{1}{p(\xi)(1+|\xi|^2)^k} e^{ix\cdot\xi}\, d\xi$$

$$= \frac{1}{(2\pi)^d} \lim_{R\to\infty} \int_{\mathbb{R}^d} \zeta_R(\xi)\, \frac{1}{p(\xi)} e^{ix\cdot\xi}\, d\xi,$$

$$D^\alpha G(x) = \frac{1}{(2\pi)^d} \lim_{R\to\infty} \int_{\mathbb{R}^d} \zeta_R(\xi)\, \frac{i^{|\alpha|}\xi^\alpha}{p(\xi)} e^{ix\cdot\xi}\, d\xi$$

$$= \frac{1}{(2\pi)^d} \frac{1}{|x|^{2r}} \lim_{R\to\infty} \int_{\mathbb{R}^d} \zeta_R(\xi)\, \frac{i^{|\alpha|}\xi^\alpha}{p(\xi)} (-\Delta_\xi)^r e^{ix\cdot\xi}\, d\xi.$$

After this we integrate by parts.

EXERCISE 1.3.4. An operator $L = \sum_{|\alpha|=m} a^\alpha D^\alpha$ we call *homogeneous elliptic* of the mth order if $\sum_{|\alpha|=m} a^\alpha \xi^\alpha \ne 0$ provided $\xi \in \mathbb{R}^d \setminus \{0\}$. For these operators Theorem 1.2.1 is not true. However, prove that if L is a mth order homogeneous elliptic operator, then for any $f \in C_0^\infty(\mathbb{R}^d)$ there is a sequence of functions $u_n \in C_b^\infty(\mathbb{R}^d)$ such that $D^\alpha L u_n \to D^\alpha f$ uniformly in $\mathbb{R}^d$ for any multi–index α.

EXERCISE 1.3.5. Prove that if L is a mth order homogeneous elliptic operator and $m = 1$, then $d \leq 2$. Also prove that if $m = 1, d = 2$, then there is a linear change of coordinates such that the operator L takes the form $c\partial/\partial\bar{z}$ where the constant $c \neq 0$ and *the Cauchy-Riemann operator*

$$\frac{\partial}{\partial \bar{z}} := \frac{1}{2}\Big(\frac{\partial}{\partial x} + i\frac{\partial}{\partial y}\Big).$$

EXERCISE 1.3.6. Prove that if L is a mth order elliptic operator and $m = 1$, then $d = 1$.

EXERCISE 1.3.7. Prove that if the coefficients a^α are real and $d \geq 2$, then m is even.

1.4. Green's functions as usual functions

Starting with this section we assume that $m \geq 2$.

THEOREM 1.4.1. *The generalized function G is a usual (locally integrable) function. If $m > d$, then G is bounded and continuous. In the general case for $x \neq 0$ and any integer $r = 0, 1, 2, \ldots$ such that $m + 2r > d$ we have*

$$G(x) = \frac{1}{(2\pi)^d}\frac{1}{|x|^{2r}}\int_{\mathbb{R}^d} e^{ix\cdot\xi}(-\Delta)^r[p(\xi)^{-1}]\,d\xi. \tag{1.4.1}$$

Moreover, if $m \leq d$ and $d - m$ is even, then for $x \neq 0$ and $m + 2r = d + 2$

$$G(x) = \frac{1}{(2\pi)^d}\frac{ix^k}{|x|^{2r}}\int_{\mathbb{R}^d} e^{ix\cdot\xi}(-\Delta)^{r-1}[p(\xi)^{-1}]_{\xi^k}\,d\xi. \tag{1.4.2}$$

Proof. First we prove formula (1.4.2), the right-hand side of which we denote by $g(x)$. Observe that since $m \geq 2$ and $2r = d - m + 2$, the integral in the definition of g is bounded (see Lemma 1.1.9), $|g(x)| \leq N|x|^{-(2r-1)} = N|x|^{m-2}|x|^{-(d-1)}$, and the function g is locally integrable. Also take a $\zeta \in C_0^\infty(\mathbb{R}^d)$ such that $\zeta(\xi) = 1$ for $|\xi| \leq 1$ and $\zeta(\xi) = 0$ for $|\xi| \geq 2$ and denote

$$\begin{aligned} g_R(x) &:= \frac{1}{(2\pi)^d}\frac{ix^k}{|x|^{2r}}\int_{\mathbb{R}^d} e^{ix\cdot\xi}(-\Delta)^{r-1}[\zeta_R(\xi)p^{-1}(\xi)]_{\xi^k}\,d\xi \\ &= \frac{1}{(2\pi)^d}\frac{1}{|x|^{2r}}\int_{\mathbb{R}^d} e^{ix\cdot\xi}(-\Delta)^r[\zeta_R(\xi)p^{-1}(\xi)]\,d\xi, \end{aligned}$$

where the second equality follows after integrating by parts. Owing to Remark 1.3.3, this shows that $g_R \to G$ in the sense of distributions. Furthermore, for some constants $c^{\alpha\beta}$ we have

$$\begin{aligned} g_R(x) =& \frac{1}{(2\pi)^d}\frac{ix^k}{|x|^{2r}}\int_{\mathbb{R}^d} e^{ix\cdot\xi}\zeta_R(\xi)\,(-\Delta)^{r-1}[p(\xi)^{-1}]_{\xi^k}\,d\xi \\ &+ \frac{1}{(2\pi)^d}\frac{ix^k}{|x|^{2r}}\sum_{\substack{|\alpha|+|\beta|=2r-1\\|\alpha|\geq 1}} c^{\alpha\beta}\int_{\mathbb{R}^d} e^{ix\cdot\xi}[D^\alpha\zeta_R(\xi)]\,D^\beta[p(\xi)^{-1}]\,d\xi. \end{aligned} \tag{1.4.3}$$

Here the first integral on the right is uniformly bounded in x and the first term on the right converges to $g(x)$ for any $x \neq 0$. Since its absolute value is also less than a locally integrable function $N|x|^{-2r+1}$ with a constant N independent of R, it follows that it converges to g in the sense of distributions as well. Now, to prove formula (1.4.2) it remains only to prove that the second term on the right in (1.4.3) goes to zero in the sense of distributions.

To do this use the local integrability of $|x|^{-2r+1}$ again and notice that if $|\alpha| + |\beta| = 2r - 1$ and $\alpha \neq 0$ and $R > 1$, then from the properties of ζ we infer that (cf. Lemma 1.1.9)

$$|[D^\alpha \zeta(\xi/R)]\, D^\beta[p(\xi)^{-1}]| \leq N \frac{1}{R^{|\alpha|}} \frac{1}{1 + |\xi|^{m+|\beta|}} I_{|\xi| \leq 2R},$$

where $I_{|\xi|\leq 2R}$ is the indicator function of the set $\{\xi : |\xi| \leq 2R\}$. (Generally, if A is a set, $I_A(\xi) = 1$ for $\xi \in A$ and $= 0$ for $\xi \notin A$.) Also,

$$1 + |\xi|^{m+|\beta|} \geq (2R)^{1-|\alpha|}(1 + |\xi|^{m+|\beta|+|\alpha|-1}) = (2R)^{1-|\alpha|}(1 + |\xi|^d)$$

for $|\xi| \leq 2R$ and $R > 1$. Hence

$$|[D^\alpha \zeta_R(\xi)]\, D^\beta[p(\xi)^{-1}]| \leq \frac{N}{R} \frac{1}{1 + |\xi|^d} I_{|\xi|\leq 2R},$$

which yields

$$\int_{\mathbb{R}^d} |[D^\alpha \zeta_R(\xi)] D^\beta[p^{-1}(\xi)]|\, d\xi \leq \frac{N}{R} \int_{|\xi|\leq 2R} \frac{1}{1 + |\xi|^d}\, d\xi \to 0$$

as $R \to \infty$. This proves (1.4.2). Observing that $ix^k e^{ix\cdot\xi} = (e^{ix\cdot\xi})_{\xi^k}$ and integrating by parts in (1.4.2), we also obtain (1.4.1) in the particular case under consideration. For m still less than or equal to d and $d - m$ still even, but for larger values of r we get (1.4.1) integrating by parts and using (1.3.3).

Finally, in the remaining case, in which either $m > d$ or $m \leq d$ and $d - m$ is odd, we first take the smallest integer $r_0 \geq 0$ such that $m + 2r_0 \geq d + 1$. If $r_0 = 0$ $(m > d)$, then we get (1.4.1) for $r = r_0$ just from (1.2.2). If $r_0 \geq 1$, then $m \leq d$ and $d - m$ is odd. Also $m + 2r_0 = d + 1$, $2r_0 < d$ $(m \geq 2\ !)$, the function $|x|^{-2r_0}$ is locally integrable, and we can obtain (1.4.1) for $r = r_0$ as in the first part of the proof. The extension to larger values of r is done as above. The theorem is proved.

1.5. Differentiability of Green's functions

COROLLARY 1.5.1. *The function $G(x)$ is infinitely differentiable for $x \neq 0$. In particular, $LG(x) = 0$ for $|x| \neq 0$ in usual rather than generalized sense. For any multi-index α we have $|D^\alpha G(x)| = o(|x|^{-n})$ as $|x| \to \infty$ for any $n > 0$. Furthermore, for any $x \neq 0$ and any multi-index α and integer $r = 0, 1, 2, \ldots$ such that $m + 2r > d + |\alpha|$*

$$D^\alpha G(x) = \frac{1}{(2\pi)^d} \frac{1}{|x|^{2r}} \int_{\mathbb{R}^d} e^{ix\cdot\xi}(-\Delta)^r \left[\frac{i^{|\alpha|}\xi^\alpha}{p(\xi)}\right] d\xi. \qquad (1.5.1)$$

Indeed, the first three assertions are true since r in (1.4.1) can be taken to be arbitrarily large. To prove the last one notice that $D^\alpha G(x)$ is a usual function which is even continuous outside zero. Furthermore, the limit on the right in (1.3.2) is uniform in the whole of $\mathbb{R}^d$. Indeed (cf. Lemma 1.1.9)

$$
\begin{aligned}
&\left|\int_{\mathbb{R}^d} e^{ix\cdot\xi}\zeta_R(\xi)(-\Delta)^r\left[\frac{i^{|\alpha|}\xi^\alpha}{p(\xi)}\right]d\xi - \int_{\mathbb{R}^d} e^{ix\cdot\xi}(-\Delta)^r\left[\frac{i^{|\alpha|}\xi^\alpha}{p(\xi)}\right]d\xi\right| \\
&\le \int_{\mathbb{R}^d} |1-\zeta_R(\xi)|\,\left|(-\Delta)^r\left[\frac{i^{|\alpha|}\xi^\alpha}{p(\xi)}\right]\right|d\xi \\
&\le N\int_{\mathbb{R}^d} |1-\zeta_R(\xi)|\,\frac{1}{1+|\xi|^{m+2r-|\alpha|}}\,d\xi,
\end{aligned}
$$

which tends to zero as $R\to\infty$, since $\zeta(0)=1$ and $m+2r-|\alpha|\ge d+1$. Also, in the same way as above for $|\gamma|+|\beta|=2r$ and $\gamma\ne 0$

$$
\begin{aligned}
\int_{\mathbb{R}^d} \left|[D^\gamma\zeta_R(\xi)]\,D^\beta\left[\frac{i^{|\alpha|}\xi^\alpha}{p(\xi)}\right]\right|d\xi &\le NR^{-|\gamma|}\int_{|\xi|\le 2R}\frac{1}{1+|\xi|^{m+|\beta|-|\alpha|}}\,d\xi \\
&\le NR^{-1}\int_{|\xi|\le 2R}\frac{1}{1+|\xi|^{m+|\gamma|+|\beta|-|\alpha|-1}}\,d\xi \\
&\le NR^{-1}\int_{|\xi|\le 2R}\frac{1}{1+|\xi|^{d}}\,d\xi \to 0.
\end{aligned}
$$

Thus proved, Corollary 1.5.1 shows that the function $G(x)$ is actually integrable and not only locally integrable. Therefore, formula (1.2.1) makes sense at least for any bounded (measurable) function f. If also the nth order derivatives of f are bounded and continuous, the same holds for the nth order derivatives of u. Moreover, if $n\ge m$, the formula still defines a solution of the equation $Lu=f$. To see this, take uniformly bounded functions $f_n\in C_0^\infty(\mathbb{R}^d)$ such that $f_n(x)\to f(x)$ at any x, and observe that the functions $G*f_n$ are uniformly bounded and converge to $G*f$ at any point. Therefore, in the sense of distributions $G*f_n\to G*f$, $f_n=L(G*f_n)\to L(G*f)$ and $L(G*f)=f$. We arrive at the following generalization of Theorem 1.2.1, in which by $C_b^n(\mathbb{R}^d)$ we mean the space of all functions on $\mathbb{R}^d$ which have bounded and continuous derivatives up to the nth order.

THEOREM 1.5.2. *Let $n\ge m$ be an integer. Then for any $f\in C_b^n(\mathbb{R}^d)$ formula (1.2.1) defines a function $u\in C_b^n(\mathbb{R}^d)$ such that $Lu=f$ in $\mathbb{R}^d$. In other words formula (1.2.3) holds for any $x\in\mathbb{R}^d$ and $f\in C_b^n(\mathbb{R}^d)$.*

REMARK 1.5.3. We will later prove much sharper results on solvability of the equation $Lu=f$. At this point let us only observe that actually we do not need f and its derivatives bounded in order for formula (1.2.3) to hold. This formula also holds if f and its derivatives up to and including the mth order derivatives are bounded by a polynomial. To see this it suffices to take the same ζ as above and use

$$f(x) = \lim_{R\to\infty} f(x)\zeta_R(x)$$

$$= \lim_{R\to\infty} \int_{\mathbb{R}^d} G(x-y)L(f\zeta_R)(y)\,dy = \int_{\mathbb{R}^d} G(x-y)Lf(y)\,dy,$$

the last equality being true since $L(f\zeta_R) \to Lf$ and $|L(f\zeta_R)|$ are bounded by a polynomial independent of R if $R > 1$.

EXERCISE 1.5.4. Prove that $\int_{\mathbb{R}^d} G(x)\,dx = (a^0)^{-1}$.

1.6. Some properties of solutions of $Lu = f$

Denote $B_R = \{x \in \mathbb{R}^d : |x| < R\}$, $B_R(x_0) = \{x \in \mathbb{R}^d : |x - x_0| < R\}$. Also for $n = 1, 2, ...$ define $C^n_{loc}(B_R)$ as the space of all functions u such that $u\zeta \in C^n_b(\mathbb{R}^d)$ for any $\zeta \in C^\infty_0(\mathbb{R}^d)$ such that $\zeta = 0$ outside B_R and near ∂B_R. Finally, remember that $m \geq 2$.

COROLLARY 1.6.1. *If $R > 0$ and a function $u \in C^m_{loc}(B_R)$ and Lu is infinitely differentiable in B_R (in particular, if $Lu = 0$), then u is infinitely differentiable in B_R.*

Indeed, take any $\varepsilon \in (0, R)$ and a function $\zeta \in C^\infty_0(\mathbb{R}^d)$ such that $\zeta(x) = 1$ for $|x| \leq R - \varepsilon$ and $\zeta(x) = 0$ for $|x| \geq R - \varepsilon/2$. Then by Theorem 1.5.2 for $|x| \leq R - \varepsilon$, we have

$$u(x) = u(x)\zeta(x) = \int_{\mathbb{R}^d} G(x-y)L(u\zeta)(y)\,dy = \int_{\mathbb{R}^d} G(x-y)\zeta(y)Lu(y)\,dy$$

$$+ \int_{\mathbb{R}^d} G(x-y) \sum_{\substack{|\alpha|+|\beta|\leq m \\ |\alpha|\geq 1}} c^{\alpha\beta}[D^\alpha\zeta]D^\beta u(y)\,dy = \int_{\mathbb{R}^d} G(y)\zeta(x-y)Lu(x-y)\,dy$$

$$+ \int_{|y|>R-\varepsilon} G(x-y) \sum_{\substack{|\alpha|+|\beta|\leq m \\ |\alpha|\geq 1}} c^{\alpha\beta}[D^\alpha\zeta]D^\beta u(y)\,dy, \quad (1.6.1)$$

where $c^{\alpha\beta}$ are some constants. The first integral on the right is infinitely differentiable for all x since $\zeta Lu \in C^\infty_0(\mathbb{R}^d)$. The second one is infinitely differentiable for $|x| < R - \varepsilon$ since $G(x)$ is infinitely differentiable outside zero.

THEOREM 1.6.2 (uniqueness). *If $u \in C^m_{loc}(\mathbb{R}^d)$ and $Lu = 0$ in $\mathbb{R}^d$ and $|u(x)| \leq N(1 + |x|^n)$ for some constants N, n and all x, then $u \equiv 0$.*

Proof. Take a function $\zeta \in C^\infty_0(\mathbb{R}^d)$ such that $\zeta(x) = 1$ for $|x| \leq 1$ and $\zeta(x) = 0$ for $|x| \geq 2$, and let $\zeta_R(x) = \zeta(x/R)$. Then for $R > 1$, for any multi-index α the functions $D^\alpha\zeta_R$ are bounded by a constant independent of R, and integrating by parts as in (1.6.1) we get for some constants $b^{\alpha\beta}$ that

$$|u(0)| = \Big| \int_{\mathbb{R}^d} G(-y) \sum_{\substack{|\alpha|+|\beta|\le m \\ |\alpha|\ge 1}} c^{\alpha\beta} [D^\alpha \zeta_R] D^\beta u(y)\, dy \Big|$$

$$= \Big| \sum_{\substack{|\alpha|+|\beta|\le m \\ |\alpha|\ge 1}} b^{\alpha\beta} \int_{2R>|y|>R} u(y) [D^\beta G(-y)] D^\alpha \zeta_R(y)\, dy \Big|$$

$$\le NR^n \max_{|\beta|\le m} \int_{|y|>R} |D^\beta G(-y)|\, dy,$$

which goes to zero as $R \to \infty$ by Corollary 1.5.1. Thus $u(0) = 0$. Since any other point can be considered similarly, the theorem is proved.

In the following exercise we suggest the reader generalize Corollary 1.6.1 and Theorem 1.6.2.

EXERCISE 1.6.3. Show that if u is a "bounded" distribution (so that $\int u(x-y)\phi(y)\,dy$ is bounded in x for any $\phi \in C_0^\infty(\mathbb{R}^d)$) and $Lu = 0$ in $\mathbb{R}^d$, then $u = 0$. Derive from here the uniqueness of the Green's function in the class of integrable functions.

In the following two exercises it is useful to bear in mind that if we have $u(x) = \phi(|x|)$, then $\Delta u(x) = \phi''(r) + (d-1)\phi'(r)/r$ with $r = |x|$, and that given one solution $\phi(r)$ of an equation $a\psi'' + b\psi' + c\psi = 0$, the second one can be found as $\phi\eta$, where η' satisfies $(a\phi)\zeta' + (b\phi + 2a\phi')\zeta = 0$.

EXERCISE 1.6.4. Prove that if $d = 3$ and $L = c - \Delta$ with a constant $c > 0$, then

$$G(x) = \frac{1}{4\pi|x|} e^{-\sqrt{c}|x|}.$$

EXERCISE 1.6.5 (cf. Exercise 9.1.6). Prove that if $L = c - \Delta$ with a constant $c > 0$, then

$$G(x) = \int_0^\infty \frac{1}{(4\pi t)^{d/2}} e^{-\frac{1}{4t}|x|^2 - ct}\, dt. \tag{1.6.2}$$

By comparing the results of Exercises 1.6.4 and 1.6.5 we obtain the equality

$$\int_0^\infty \frac{1}{t^{3/2}} e^{-\frac{1}{t} - t}\, dt = \sqrt{\pi} e^{-2},$$

which we propose the reader try to get differently.

EXERCISE 1.6.6 (hard). If $R > 0$ and u is a *distribution* such that Lu is infinitely differentiable in B_R (in particular, if $Lu = 0$), then u is infinitely differentiable in B_R.

1.7. Some further information on G

The following result on sharp estimates of $G(x)$ and its derivatives near zero plays an important role in many situations. It shows that if $d > m$, then the behavior of G and its derivatives near zero is the same as for the function $1/|x|^{d-m}$. It also shows that in the general case if $|\alpha| = m$, then the kernel $K(x) := D^\alpha G(x)$ satisfies

$$|K(x)| \leq \frac{N}{|x|^d}, \quad |D_j K(x)| \leq \frac{N}{|x|^{d+1}}$$

for any $j = 1, ..., d$. Such estimates are crucial in the Hölder–space theory as well as in the Sobolev–space theory.

THEOREM 1.7.1. *Let α be a multi–index such that $d + |\alpha| - m > 0$. Then*

$$|D^\alpha G(x)| \leq N \frac{1}{|x|^{d+|\alpha|-m}} . \tag{1.7.1}$$

The same estimate is true for $d + |\alpha| - m = 0$ for $|x| \leq 1/2$ if we replace the right-hand side with $N \log(1/|x|)$. Finally, if $d + |\alpha| - m < 0$, then $|D^\alpha G(x)|$ is bounded.

Proof. Notice that we need to show (1.7.1) only for small $|x|$, say $|x| \leq 1/2$. Take any integer r such that $m + 2r > d + |\alpha|$ and use Corollary 1.5.1. Also take a function $\zeta \in C_0^\infty(\mathbb{R}^d)$ such that $\zeta(x) = 1$ for $|x| \leq 1$ and $\zeta(x) = 0$ for $|x| \geq 2$, and on the basis of the identity $1 = \zeta(|x|\xi) + [1 - \zeta(|x|\xi)]$ represent the integral in (1.5.1) as a sum of two integrals, with the first integral containing $\zeta(|x|\xi)$ and the second one containing $1 - \zeta(|x|\xi)$. Finally, in the first integral integrate by parts, bringing all derivatives (with respect to ξ) from $i^{|\alpha|}\xi^\alpha p^{-1}$ on $\zeta(|x|\xi)\exp(ix \cdot \xi)$, and notice that

$$|\Delta_\xi^r [\zeta(|x|\xi)\exp(ix \cdot \xi)]| \leq N|x|^{2r} I_{|x|\cdot|\xi|\leq 2}.$$

For $d + |\alpha| - m \geq 1$ this yields

$$|D^\alpha G(x)| \leq \frac{1}{(2\pi)^d} \int_{|\xi|\leq 2/|x|} \frac{|\xi^\alpha|}{|p(\xi)|} \, d\xi + \frac{N}{|x|^{2r}} \int_{|\xi|>1/|x|} \frac{1}{|\xi|^{m-|\alpha|+2r}} \, d\xi$$

$$\leq N \int_0^{2/|x|} \frac{1}{\rho^{m-|\alpha|}} \rho^{d-1} \, d\rho + \frac{N}{|x|^{2r}} \int_{1/|x|}^\infty \frac{1}{\rho^{m-|\alpha|+2r}} \rho^{d-1} \, d\rho = \frac{N}{|x|^{d+|\alpha|-m}} .$$

If $d + |\alpha| - m = 0$, we get the result by noticing that

$$D^\alpha G(x) = -\int_1^{1/|x|} \frac{\partial}{\partial\rho} D^\alpha G(\rho x) \, d\rho + D^\alpha G(\frac{x}{|x|}).$$

Finally if $d + |\alpha| - m < 0$, it suffices to refer to Corollary 1.5.1 with $r = 0$. The theorem is proved.

In some applications the following result is useful, which along with Exercise 1.7.4 is also discussed in Exercise 3.5.4 (this result will not be used later).

THEOREM 1.7.2. *$G(x)$ is real–analytic for $x \neq 0$.*

Proof. It suffices to show that for k such that $2k+m>d$ the function $G'(x)$ from Corollary 1.3.2 is real–analytic for $x\neq 0$. Since G' is a Green's function for $(1-\Delta)^k L$ and $2k+m>d$, we may as well assume that m is as large as we like, for instance, $m>d+1$.

In that case by the same Corollary 1.3.2 by denoting $x=|x|\theta$ we obtain

$$G(x)=\frac{1}{(2\pi)^d}\frac{1}{|x|^d}\int_{\mathbb{R}^d}\frac{1}{p(\xi|x|^{-1})}e^{i\theta\cdot\xi}\,d\xi. \tag{1.7.2}$$

Next, fix a point $x_0\neq 0$ and $\varepsilon\in(0,1)$ and a unit $\eta\notin B_\varepsilon(\theta_0)\cup B_\varepsilon(-\theta_0)$ where $\theta_0=x_0|x_0|^{-1}$. For each $\theta\in B_\varepsilon(\theta_0)$ define a unique orthogonal transformation $T_{\eta,\theta}$ of $\mathbb{R}^d$ which leaves all vectors orthogonal to the plane Span (η,θ) fixed, rotates the plane in itself, and takes η into θ.

EXERCISE* 1.7.3. Find a formula for $T_{\eta,\theta}$ and show that it defines $T_{\eta,\theta}$ as an analytic function of θ when $|\theta-\theta_0|<\varepsilon$.

After this we replace ξ in (1.7.2) by $T_{\eta\theta}\xi$. Since

$$\theta\cdot T_{\eta\theta}\xi=T_{\eta\theta}\eta\cdot T_{\eta\theta}\xi=\eta\cdot\xi,$$

it follows that for $|x|=\rho$

$$G(x)=\frac{1}{(2\pi)^d}\frac{1}{\rho^d}\int_{\mathbb{R}^d}\frac{1}{p(T_{\eta\theta}\xi\rho^{-1})}e^{i\eta\cdot\xi}\,d\xi=:g(\theta,\rho)$$

(remember that η is fixed). It turns out that the function g is an analytic function of variables (θ,ρ) for $|\theta-\theta_0|<\varepsilon$ and $|\rho-|x_0||<\varepsilon$ if ε is small enough. Indeed, by Lemma 1.1.7 the relationships

$$|p(T_{\eta\theta_0}\xi|x_0|^{-1})|\geq\kappa(1+|T_{\eta\theta_0}\xi|x_0|^{-1}|^m)=\kappa(1+|\xi|^m/|x_0|^m)$$

imply that for ε small enough we have

$$|p(T_{\eta\theta}\xi\rho^{-1})|\geq(\kappa/2)(1+|\xi|^m/|x_0|^m)$$

for all $\xi\in\mathbb{R}^d$ and complex ρ,θ if $|\theta-\theta_0|<\varepsilon$ and $|\rho-|x_0||<\varepsilon$. This allows us to differentiate under the integral sign in the definition of the function g, so that g is analytic indeed. An analytic function is also real–analytic, and the theorem is proved.

EXERCISE 1.7.4 (hard). Show that if for a distribution u we have $Lu=0$ in B_R, then u is real–analytic in B_R.

REMARK 1.7.5. Everything above can be easily extended to systems when a^α are square matrices with n rows and n columns, so that the equation $Lu=f$ is an mth-order system for a vector u with n components. In that case we deal with matrix–valued characteristic polynomials, and the definition of the ellipticity becomes $|\det p(\xi)|\geq\kappa(1+|\xi|^{mn})\quad\forall\xi\in\mathbb{R}^d$, so that again $|p^{-1}(\xi)|\leq N(1+|\xi|^m)^{-1}$.

1.8. Hints to exercises

1.2.2. Try $\exp(\xi \cdot x)$.
1.3.4. Take the operator $M_\varepsilon = (L\bar{L}^*)^k + \varepsilon$, where the integer k is so large that $2mk > d$ and $\varepsilon > 0$. Observe that the characteristic polynomial of M_n is $|p(\xi)|^{2k} + \varepsilon$, where $p(\xi) = \sum_{|\alpha|=m} a^\alpha i^{|\alpha|} \xi^\alpha$. Next, define $v_\varepsilon \in C_b^\infty(\mathbb{R}^d)$ so that $M_\varepsilon v_\varepsilon = f$, and prove that for any $r \geq 0$

$$\int_{\mathbb{R}^d} |\xi|^r |\tilde{f}(\xi) - |p(\xi)|^{2k} \tilde{v}_\varepsilon(\xi)| \, d\xi \to 0$$

as $\varepsilon \downarrow 0$. After this take $u_n = \bar{L}^*(L\bar{L}^*)^{k-1} v_{1/n}$.
1.3.7. Take two linearly independent vectors ξ, η and consider $p(\xi + t\eta)$ as a polynomial in real t.
1.6.3. Consider the functions $v(x) = \int u(x-y)\phi(x)\, dy$.
1.6.4. From uniqueness of G conclude that $G(x) = G(|x|)$. After this find all vanishing at infinity solutions depending only on $|x|$ of the equation $Lu = 0$ outside the origin and apply Exercise 1.5.4.
1.6.5. Proceed as in Exercise 1.6.4 and show that any radial solution of $Lu = 0$ not proportional to G tends to infinity as $|x| \to \infty$.

One also might like to check (1.2.3) for all $f \in C_0^\infty(\mathbb{R}^d)$ by passing to the Fourier transforms and applying Parseval's identity.
1.6.6. Consider $u * \phi_n$, where ϕ_n approximate δ_0, and apply (1.6.1).

CHAPTER 2

Laplace's Equation

There is certain danger in dealing only with the general theory. One can miss really beautiful and extremely instructive ideas, the major part of which was known in the nineteenth century. We present some of them in this chapter. We consider only real–valued solutions in domains Ω which are sufficiently regular to allow us integration by parts. Denote by $C^n(\bar{\Omega})\,(C^n(\Omega))$ the space of all functions having continuous and bounded derivatives in $\bar{\Omega}$ (respectively, in Ω) up to the nth order. Of course, $C(\bar{\Omega}) = C^0(\bar{\Omega})$ and $C(\Omega) = C^0(\Omega)$.

2.1. Green's identities

Let Ω be a sufficiently regular bounded domain. For $u, v \in C^2(\bar{\Omega})$ we have the following Green's identities

$$\begin{aligned} \int_\Omega v\Delta u\,dx &= -\int_\Omega v_{x^i}u_{x^i}\,dx + \int_{\partial\Omega} v\frac{\partial u}{\partial n}\,dS, \\ \int_\Omega v\Delta u\,dx &= \int_\Omega u\Delta v\,dx + \int_{\partial\Omega}\Big(v\frac{\partial u}{\partial n} - u\frac{\partial v}{\partial n}\Big)\,dS, \end{aligned} \tag{2.1.1}$$

where $\partial/(\partial n)$ indicates differentiation in the direction of the *exterior* normal to $\partial\Omega$.

A special case of interest is $v = u$. We find the *energy identity*

$$\int_\Omega |u_x|^2\,dx + \int_\Omega u\Delta u\,dx = \int_{\partial\Omega} u\frac{\partial u}{\partial n}\,dS.$$

If here $\Delta u = 0$ in Ω and either $u = 0$ or $\partial u/\partial n = 0$ on $\partial\Omega$, it follows that $|u_x| = 0$ in Ω and $u = \text{const}$ in $\bar{\Omega}$. This observation leads to uniqueness theorems for two of the standard problems of the theory of second–order elliptic equations:

The *Dirichlet problem*: Find u in Ω from prescribed values of Δu in Ω and of u on $\partial\Omega$.

The *Neumann problem*: Find u in Ω from prescribed values of Δu in Ω and of $\partial u/\partial n$ on $\partial\Omega$.

Thus: *A solution $u \in C^2(\bar{\Omega})$ of the Dirichlet problem is determined uniquely. A solution $u \in C^2(\bar{\Omega})$ of the Neumann problem is determined uniquely within an additive constant.*

In the case of the Neumann problem, to be sure that solutions exist we need more than only certain regularity of data. For example, after taking $v \equiv 1$ in (2.1.1) we see that

$$\int_\Omega \Delta u\,dx = \int_{\partial\Omega} \frac{\partial u}{\partial n}\,dS,$$

which is a nontrivial condition on values of Δu in Ω and $\partial u/\partial n$ on $\partial\Omega$.

DEFINITION 2.1.1. Given a domain $\Omega \subset \mathbb{R}^d$ and a twice continuously differentiable function u defined in Ω, the function u is called *harmonic* in Ω if $\Delta u = 0$ in Ω.

We will later see that harmonic functions are infinitely differentiable. For $d = 2$ this can be obtained from the theory of analytic functions on the basis of the following exercise.

EXERCISE 2.1.2. Let x, y denote coordinates in the plane, and let $u(x, y)$ be a harmonic function in a simply connected domain Ω. Define

$$v(x,y) = \int_{(x_0,y_0)}^{(x,y)} (u_x\, dy - u_y\, dx)$$

and prove that $v(x, y)$ is a harmonic function *conjugate* to u, that is, $u_x = v_y$, $u_y = -v_x$, so that $f = u+iv$ satisfies the Cauchy–Riemann conditions and therefore is analytic in Ω.

EXERCISE 2.1.3. Let $d = 2$ and let u, v be conjugate harmonic functions continuously differentiable in the closure of a simply connected domain Ω whose boundary is smooth. Show that on the boundary curve $\partial\Omega$

$$\frac{\partial u}{\partial n} = \frac{\partial v}{\partial s}, \quad \frac{\partial v}{\partial n} = -\frac{\partial u}{\partial s},$$

where ∂n denotes differentiation in the direction of the exterior normal and ∂s differentiation in the counterclockwise tangential direction. Show how these relations can be used to reduce the Neumann problem for u to the Dirichlet problem for v and conversely.

EXERCISE 2.1.4. Let $d = 2$ and $\Omega = \{(x, y) : x > 0\}$. By considering z^δ prove that if $\delta \in (0, 1)$, then there exists a harmonic nonnegative function in Ω which is continuous in $\bar{\Omega}$ and equals $|y|^\delta$ on $\partial\Omega$.

EXERCISE 2.1.5. Let u satisfy $\Delta u(x) = f(x)$. Take an orthogonal matrix p and define $v(x) = u(px)$. Show that $\Delta v(x) = f(px)$.

2.2. The Poisson formula

The spherical symmetry of the operator Δ (cf. Exercise 2.1.5) suggests that there are solutions of the equation $\Delta u = 0$ which depend only on $r = |x|$. For functions $v(x) = \psi(r)$ we have

$$\Delta v(x) = \psi''(r) + \frac{d-1}{r}\psi'(r), \quad r > 0.$$

Therefore we easily find that the following function solves $\Delta v = 0$ apart from the origin

$$v(x) = \psi(r) = \begin{cases} \frac{1}{(2-d)\omega_d} r^{2-d} & d \geq 3, \\ \frac{1}{2\pi} \log r & d = 2, \\ -\frac{1}{2} r & d = 1, \end{cases}$$

where ω_d is the surface measure of the unit sphere in $\mathbb{R}^d$ ($\omega_1 = 2$). Actually we can multiply the function ψ by any constant. The particular choice of the constant is prompted by the following theorem.

THEOREM 2.2.1. *For $x, y \in \mathbb{R}^d$ let $K(x,y) = \psi(|x-y|)$. Then $\Delta_y K = \delta_x$ in the sense of distributions, that is,*

$$\int_{\mathbb{R}^d} K(x,y)\Delta\phi(y)\,dy = \phi(x) \quad \forall \phi \in C_0^\infty(\mathbb{R}^d), x \in \mathbb{R}^d.$$

Moreover, if u is a bounded function with compact support and u is of class $C^2(B_r(x_0))$ for some $r > 0$ and x_0, then the Poisson formula holds

$$\Delta \int_{\mathbb{R}^d} u(y)K(x,y)\,dy|_{x=x_0} = u(x_0).$$

In particular, we claim that the integral involved is twice continuously differentiable in a certain neighborhood of x_0.

Proof. From the formula

$$\int_{\mathbb{R}^d} K(x,y)\Delta\phi(y)\,dy = \int_{\mathbb{R}^d} \psi(|y|)\Delta_x\phi(x-y)\,dy = \Delta \int_{\mathbb{R}^d} K(x,y)\phi(y)\,dy$$

we see that it suffices to prove the second assertion. In the proof of the latter we may take $x_0 = 0$. Also derivatives of $K(x,y)$ are bounded if the arguments x, y are not close and $\Delta_x K(x,y) = 0$. Therefore, if we take an infinitely differentiable function ζ with support in $B_r(x_0)$ such that $\zeta = 1$ in $B_{r/2}(x_0)$, then in the formula

$$\int_{\mathbb{R}^d} u(y)K(x,y)\,dy = \int_{\mathbb{R}^d} \zeta(y)u(y)K(x,y)\,dy + \int_{\mathbb{R}^d} (1-\zeta(y))u(y)K(x,y)\,dy$$

the last term is a harmonic function in $B_{r/2}(x_0)$. This easily implies that it remains only to consider u which is zero outside B_r. In this case, as above,

$$\Delta \int_{\mathbb{R}^d} u(y)K(x,y)\,dy|_{x=0} = \int_{\mathbb{R}^d} \psi(|y|)\Delta u(y)\,dy.$$

Next we use Green's identity

$$\int_{\mathbb{R}^d} \psi(|y|)\Delta u(y)\,dy = \lim_{\rho\downarrow 0} \int_{B_\rho^c} \psi(|y|)\Delta u(y)\,dy = \lim_{\rho\downarrow 0} \int_{\partial B_\rho} \left(\psi\frac{\partial u}{\partial n} - u\frac{\partial \psi}{\partial n}\right) dS.$$

Here

$$|\int_{\partial B_\rho} \psi\frac{\partial u}{\partial n}\,dS| \le \omega_d\psi(\rho)\rho^{d-1}\max|u_x| \to 0,$$

$$-\int_{\partial B_\rho} u\frac{\partial\psi}{\partial n}\,dS = \frac{1}{\omega_d\rho^{d-1}}\int_{\partial B_\rho} u\,dS$$
$$= \frac{1}{\omega_d}\int_{\partial B_1} u(x\rho)\,dS \to u(0)\frac{1}{\omega_d}\int_{\partial B_1} dS = u(0).$$

The theorem is proved.

By taking $u = fI_\Omega$ we get the following statement which shows that in order to solve the equation $\Delta w = f$ one may take solutions $K(x, y)$ of the equation $\Delta_x K = \delta_y$ and "sum" them up with respect to y after multiplying by $f(y)$. Because of this property the function K is often called *the fundamental solution* of the equation $\Delta w = f$.

COROLLARY 2.2.2. *Let Ω be a bounded domain, $f \in C^2(\Omega)$. Then the formula*

$$w(x) = \int_\Omega K(x, y) f(y)\, dy$$

defines a bounded continuous function in $\mathbb{R}^d$ which satisfies Poisson's equation $\Delta w = f$ in Ω.

This corollary shows that instead of solving the equation $\Delta u = f$ in Ω with boundary data $u = g$ on $\partial\Omega$, one needs only to find a harmonic function h in Ω which equals $g - w$ on $\partial\Omega$. Indeed, in that case $u = h + w$ will give a solution to the initial problem.

2.3. Green's functions in domains

The fundamental solution helps us to investigate the Dirichlet problem: Given two functions f, g in $\bar{\Omega}$, find $u \in C^2_{loc}(\Omega) \cap C(\bar{\Omega})$ such that

$$\Delta u = f \quad \text{in} \quad \Omega, \quad u(x) = g(x) \quad \text{for} \quad x \in \partial\Omega. \tag{2.3.1}$$

THEOREM 2.3.1. *Let Ω be a regular bounded domain. Assume that for any $x \in \Omega$ there exists a function $h(x, \cdot) \in C^2(\bar{\Omega})$ such that*

$$\Delta_y h(x, y) = 0 \quad in \quad \Omega, \quad h(x, y) = K(x, y) \quad for \quad y \in \partial\Omega.$$

Define the Green's function

$$G(x, y) = K(x, y) - h(x, y)$$

so that, in particular, $\Delta_y G(x, y) = 0$ in $\Omega \setminus \{x\}$ and $G(x, y) = 0$ for $y \in \partial\Omega$. Then for any $C^2(\bar{\Omega})$-solution u of the Dirichlet problem (2.3.1) and $x \in \Omega$ we have

$$u(x) = \int_\Omega G(x, y) f(y)\, dy + \int_{\partial\Omega} H(x, y) g(y)\, dS_y, \tag{2.3.2}$$

where

$$H(x, y) := \frac{\partial}{\partial n_y} G(x, y), \quad x \in \Omega, y \in \partial\Omega$$

is the so-called Poisson kernel.

Proof. The case $d = 1$ is left to the reader, who is also asked to check that actually the proof below is valid for $d = 1$ as well. If $d \geq 2$, let $x \in \Omega$ and $\Omega_\rho = \Omega \setminus B_\rho(x)$. Since $\Delta_y G(x, y) = 0$, by Green's identity we get

$$\int_\Omega G(x, y) f(y)\, dy = \lim_{\rho \downarrow 0} \int_{\Omega_\rho} G(x, y) \Delta u(y)\, dy,$$

$$\int_{\Omega_\rho} G(x,y)\Delta u(y)\,dy = \int_{\partial\Omega_\rho} [G(x,y)\frac{\partial u}{\partial n_y} - u(y)\frac{\partial}{\partial n_y}G(x,y)]\,dS_y$$

$$= \int_{\partial B_\rho} [G(x,y)\frac{\partial u}{\partial n_y}(y) - u(y)\frac{\partial}{\partial n_y}G(x,y)]\,dS_y - \int_{\partial\Omega} g(y)\frac{\partial}{\partial n_y}G(x,y)\,dS_y.$$

Here as in the proof of Theorem 2.2.1

$$\int_{\partial B_\rho} G(x,y)\frac{\partial u}{\partial n_y}(y)\,dS_y \to 0,$$

$$\int_{\partial B_\rho} u(y)\frac{\partial}{\partial n_y}G(x,y)\,dS_y = \int_{\partial B_\rho} u(y)\frac{\partial}{\partial n_y}K(x,y)\,dS_y - \int_{\partial B_\rho} u(y)\frac{\partial}{\partial n_y}h(x,y)\,dS_y,$$

$$\int_{\partial B_\rho} u(y)\frac{\partial}{\partial n_y}K(x,y)\,dS_y \to u(x), \quad \int_{\partial B_\rho} u(y)\frac{\partial}{\partial n_y}h(x,y)\,dS_y \to 0.$$

The theorem is proved.

REMARK 2.3.2. In the proof we did not use the fact that $h(x,y)$ is given for all $x \in \Omega$. Therefore, if we manage to find $h(x_0,y)$ only for some $x_0 \in \Omega$, formula (2.3.2) will hold for the same $x = x_0$.

In particular, if $\Omega = B_R$ and $x_0 = 0$, then $K(0,y) = \psi(R)$ for $|y| = R$ and, obviously, $h(0,y) \equiv \psi(R)$ is a harmonic function equal to K on the boundary. Therefore, for any $C^2(\bar{B}_R)$–harmonic function u we have

$$u(0) = \frac{1}{\omega_d R^{d-1}} \int_{|y|=R} u(y)\,dS. \tag{2.3.3}$$

Generally for x such that the ball $B_R(x)$ lies in a domain where u is harmonic we have *Gauss's law of the arithmetic mean*

$$u(x) = \frac{1}{\omega_d R^{d-1}} \int_{|x-y|=R} u(y)\,dS_y. \tag{2.3.4}$$

EXERCISE 2.3.3. Let $d = 2$ and $\Omega = \{z = (x,y) : y > 0\}$ be a halfplane. For $z = (x,y)$ define by reflection $z^* = (x,-y)$. Show that $G(z_1,z_2) = K(z_1,z_2) - K(z_1,z_2^*)$ is a Green's function for Ω. Find the Poisson kernel and prove that for any bounded and continuous function $f(x)$ the formula

$$u(x,y) = \frac{1}{\pi}\int_{-\infty}^{\infty} \frac{yf(t)}{(x-t)^2 + y^2}\,dt$$

defines a harmonic function in Ω and $u(x,y) \to f(x)$ as $y \downarrow 0$.

EXERCISE 2.3.4. Assume that u is twice continuously differentiable in a neighborhood of the origin and that the right–hand side of (2.3.3) is independent of R for all small R. Prove that $\Delta u(0) = 0$.

EXERCISE 2.3.5. Assume that u is a bounded (and measurable) function in Ω for which (2.3.4) holds whenever $B_R(x) \subset \Omega$. Prove that u is infinitely differentiable in Ω and u is harmonic in Ω.

2.4. The Green's function and the Poisson kernel in a ball

Theorem 2.3.1 makes it plausible that if we manage to find the Green's function and the Poisson kernel for Ω, then formula (2.3.2) will actually give us a solution of the Dirichlet problem. This way of finding solutions explicitly and hence investigating their properties is more or less completely abandoned in modern theories. The main reason is that in the more or less general situation of equations with variable coefficients, one cannot hope to find anything explicitly, even though very often one gets rather complete information about solutions by other means.

It turns out that for some domains the Green's function and the Poisson kernel can indeed be found explicitly. One such situation is described in Exercise 2.3.3. Here we let $\Omega = B_R$.

Let us use the following result of simple computations: If $d \geq 2$ and $\Delta u = 0$ in a domain Ω', then

$$\Delta\Big(|x|^{2-d} u\Big(\frac{x}{|x|^2}\Big)\Big) = 0$$

for x such that $x/|x|^2 \in \Omega'$. Denote $y^* = R^2 y/|y|^2$. It follows that the function

$$K^*(x,y) := K(x,y^*)|y|^{2-d}R^{d-2}$$

is harmonic with respect to y for $y \neq x^*$. It is remarkable that x^* lies outside the ball B_R whenever x is inside, and $K(x,y) = K^*(x,y)$ if $|y| = R$. Therefore, by Theorem 2.3.1 the formulas

$$G(x,y) = K(x,y) - K^*(x,y), \quad H(x,y) = \frac{1}{R\omega_d}\frac{R^2 - |x|^2}{|x-y|^d}$$

give us the Green's function and the Poisson kernel for B_R. Interestingly enough these formulas are also true for $d = 1$.

In particular, if $u \in C^2(\bar{B}_R)$ and u is harmonic in B_R, then *the Poisson integral formula* holds: for any $x \in B_R$

$$u(x) = \int_{|y|=R} H(x,y)u(y)\,dS_y. \tag{2.4.1}$$

Natural questions arise: given a continuous function g and u is defined in B_R by

$$u(x) = \int_{|y|=R} H(x,y)g(y)\,dS_y, \tag{2.4.2}$$

is u a harmonic function in B_R and does it take the boundary values g on ∂B_R? The answer to both questions is "yes". In the case of the first question one can just check that $\Delta u = 0$ in B_R by an easy computation. The proof that $u(x) \to g(y)$ if $B_R \ni x \to y \in \partial B_R$ is also elementary but a little bit more technical. We will get the proof from a different argument later (see Remark 2.7.2 below).

There are many other methods that can be used to find the Green's function and the Poisson kernel for particular domains. For the case of half spaces we discuss them in Chapter 5. If $d = 2$ and $R^2 = \{(x,y)\}$, then one can apply the theory of analytic functions. To make the explanation shorter, let us introduce the Cauchy-Riemann operators

$$\frac{\partial}{\partial \bar{z}} := \frac{1}{2}\Big(\frac{\partial}{\partial x} + i\frac{\partial}{\partial y}\Big), \quad \frac{\partial}{\partial z} := \frac{1}{2}\Big(\frac{\partial}{\partial x} - i\frac{\partial}{\partial y}\Big)$$

(they arise naturally if one notices that $x = (z + \bar{z})/2$, $y = (z - \bar{z})/(2i)$ so that formally

$$\frac{\partial u}{\partial \bar{z}} = \frac{\partial u}{\partial x}\frac{\partial x}{\partial \bar{z}} + \frac{\partial u}{\partial y}\frac{\partial y}{\partial \bar{z}}\ ).$$

It turns out that the analyticity of a function u means that

$$u_{\bar{z}} = \frac{\partial u}{\partial \bar{z}} = 0,$$

so that in the power series expansion of $u(x, y)$ there are no terms with $\bar{z}$. Also,

$$\Delta u = 4\frac{\partial^2}{\partial z \partial \bar{z}}\, u,$$

which, in particular, implies that real and imaginary parts of any analytic function are (real-valued) harmonic functions. Now, if $\Delta u = f$ in a domain $\Omega \subset R^2$ and F is an analytic mapping $\Omega' \to \Omega$, then in Ω'

$$\frac{1}{4}\Delta[u(F(z))] = \frac{\partial^2}{\partial z \partial \bar{z}}[u(F(z))] = \frac{\partial}{\partial z}[u_F \cdot 0 + u_{\bar{F}}\bar{F}_{\bar{z}}]$$
$$= \frac{\partial}{\partial z}\{u_{\bar{F}}\overline{F_z}\} = u_{\bar{F}F}F_z\overline{F_z} = |F_z|^2\frac{1}{4}\{\Delta u\}(F(z)).$$

This shows that if we have a method of solving Poisson's equation in Ω and F is not too bad (say, $|F_z| > 0$), then we can solve this equation in Ω' too. In particular, we see that analytic mappings bring harmonic functions into harmonic functions.

2.5. Some properties of harmonic functions

The Poisson integral formula (2.4.1) is very useful in investigating properties of harmonic functions.

EXERCISE* 2.5.1. By using (2.4.1) prove that harmonic functions are infinitely differentiable.

Upon differentiating (2.4.1) we easily get that if $u \in C^2(\bar{B}_R(x_0))$ and u is harmonic in $B_R(x_0)$, then

$$|u_x(x_0)| \le \frac{d}{R}\max_{B_R(x_0)}|u|. \tag{2.5.1}$$

By Exercise 2.5.1 the function u is infinitely differentiable in $B_R(x_0)$, and by differentiating the equation $\Delta u = 0$ we see that, for any $i = 1, ..., d$, the function u_{x^i} is harmonic and $u_{x^i} \in C^2(\bar{B}_{R/2}(x_0))$. Therefore,

$$|u_{x^i x}(x_0)| \le \frac{d}{R/2}\max_{B_{R/2}(x_0)}|u_{x^i}|.$$

Also any point $x \in B_{R/2}(x_0)$ is the center of a ball of radius $R/2$ lying in $B_R(x_0)$. Therefore, from (2.5.1) it follows that

$$\max_{B_{R/2}(x_0)} |u_{x^i}| \le \frac{d}{R/2} \max_{B_R(x_0)} |u|, \quad |u_{x^i x}(x_0)| \le \left(\frac{2d}{R}\right)^2 \max_{B_R(x_0)} |u|.$$

By induction we get the following result.

THEOREM 2.5.2. *Let Ω be a domain and $u \in C^2_{loc}(\Omega) \cap C(\bar{\Omega})$ be a harmonic function in Ω. Then u is infinitely differentiable in Ω and for any multi–index α and any $x \in \Omega$ we have*

$$|D^\alpha u(x)| \le \left[\frac{|\alpha| d}{\operatorname{dist}(x, \partial\Omega)}\right]^{|\alpha|} \sup_\Omega |u|.$$

COROLLARY 2.5.3. *Let Ω be a domain and u_n be a uniformly bounded sequence of harmonic functions in Ω. Then for any multi–index α the family $D^\alpha u_n$ is uniformly bounded and equicontinuous on any compact set $\Gamma \subset \Omega$. Furthermore, if $u_n(x) \to u(x)$ at any point $x \in \Omega$, then u is harmonic in Ω.*

COROLLARY 2.5.4. *If u is a harmonic function in $\mathbb{R}^d$ and $|u(x)| \le N(1+|x|^n)$ for some constants N, n and all x, then u is a polynomial of degree less than or equal to n.*

Indeed, by Theorem 2.5.2 for $\Omega = B_R$ and $R \to \infty$ we find $D^\alpha u \equiv 0$ whenever $|\alpha| > n$.

Next we apply the Poisson integral formula (2.4.1) in proving the following *Harnack's inequality.*

THEOREM 2.5.5. *If u is nonnegative, harmonic in B_R and continuous in $\bar{B}_R$, then for any two points $x_1, x_2 \in B_{R/2}$ it holds that*

$$u(x_1) \le c u(x_2),$$

where the constant $c > 0$ is independent of u and R.

To prove the theorem it suffices to reduce the whole situation to the case $R = 1$ by dilations, when one takes $v(x) := u(Rx)$ instead of u, and then observe that the Poisson kernel $H(x, y)$ is bounded and bounded away from zero once $|x| \le 1/2$ and $|y| = 1$, so that by formula (2.4.1) for $x \in B_{1/2}$

$$c \int_{\partial B_1} u\, dS \le u(x) \le c^{-1} \int_{\partial B_1} u\, dS,$$

where c is a constant independent of u.

COROLLARY 2.5.6. *If we are given a sequence of nonnegative functions which are harmonic in B_R and uniformly bounded at a point $x_0 \in B_R$, then the functions are uniformly bounded in any B_r with $r < R$.*

Harnack's inequality is also true for solutions of nondegenerate second–order elliptic equations with variable coefficients. In such situations it is an extremely deep and hard result.

The following is Liouville's theorem.

THEOREM 2.5.7. *Any nonnegative function which is harmonic in the whole space $\mathbb{R}^d$ is constant.*

Indeed, it follows from Harnack's inequality with $R \to \infty$ that $u(x_1) \le cu(0)$, so that the nonnegative function u is bounded, and we can refer to Corollary 2.5.4.

This way of proving depends however on the interior gradient estimate, which is not available for general second–order elliptic operators. That is why we give another proof too.

If we subtract the constant $\inf u$ from u we get a nonnegative harmonic function v such that $\inf v = 0$. Next take a sequence x_n such that $v(x_n) \to 0$. Then from Harnack's inequality for any x we get $v(x) \le cv(x_n)$ and $v(x) \le c \lim v(x_n) = 0$, so that $v(x) \equiv 0$, and we have the result again.

2.6. The maximum principle

Here we present one of the most powerful tools in the theory of second–order elliptic equations.

THEOREM 2.6.1 (the maximum principle). *Let Ω be a bounded domain and $u \in C^2_{loc}(\Omega) \cap C(\bar{\Omega})$ be a subharmonic function in Ω, that is, $\Delta u \ge 0$ in Ω. Then*

$$\max_{\bar{\Omega}} u = \max_{\partial\Omega} u. \tag{2.6.1}$$

Proof. First assume that $\Delta u > 0$ in Ω and let x_0 be a point at which u takes its maximum value over $\bar{\Omega}$. If we assume that $x_0 \in \Omega$, then at this point $u_{x^i x^i} \le 0$, which is impossible since $\Delta u > 0$. This implies that $x_0 \notin \Omega$ and $x_0 \in \partial\Omega$. Therefore, (2.6.1) holds. In the general case for any constant $\varepsilon > 0$ we have $\Delta(u + \varepsilon|x|^2) \ge 2\varepsilon d > 0$, so that for any $x \in \bar{\Omega}$,

$$u(x) \le u(x) + \varepsilon|x|^2 \le \max_{\partial\Omega}[u(y) + \varepsilon|y|^2] \le \max_{\partial\Omega} u + N\varepsilon,$$

where N is independent of ε. Letting $\varepsilon \downarrow 0$, we see that the left-hand side in (2.6.1) is less than its right-hand side. Since the opposite inequality is obvious, the theorem is proved.

COROLLARY 2.6.2. *If, in addition, u is harmonic in Ω, then*

$$\max_{\bar{\Omega}} |u| = \max_{\partial\Omega} |u|.$$

COROLLARY 2.6.3. *There can be only one solution of class $C^2_{loc}(\Omega) \cap C(\bar{\Omega})$ to the Dirichlet problem in Ω.*

The new information in this corollary is that the function u is not supposed to belong to $C^2(\bar{\Omega})$.

EXERCISE* 2.6.4. Observe that the function $v = R^2 - |x|^2$ satisfies $\Delta v = -2d$. By using this prove that if $w \in C^2_{loc}(B_R) \cap C(\bar{B}_R)$ and $\Delta w = f$ in B_R, then $w \le v \sup(f_-)/(2d) + \max_{\partial\Omega}(w_+) \le R^2 \sup(f_-)/(2d) + \max_{\partial\Omega}(w_+)$, where $a_+ = \max(a, 0)$, $a_- = \max(-a, 0)$. Also prove that $|w| \le R^2 \sup|f|/(2d) + \max_{\partial\Omega}|w|$.

EXERCISE 2.6.5. Prove that if u is a harmonic function in a domain Ω and there exists a point $x_0 \in \Omega$ such that $u(x_0) = \max_{\bar{\Omega}} u$, then $u = \text{const}$.

EXERCISE 2.6.6. Prove that if the function $h(x, y)$ in Theorem 2.3.1 exists, then it is continuous in $\Omega \times \Omega$.

If Ω is unbounded, the natural reformulation of Corollary 2.6.2,

$$\sup_{\bar{\Omega}} |u| = \sup_{\partial\Omega} |u|, \tag{2.6.2}$$

is not true in general. But it is true if the complement of Ω is "large enough". In connection with this do the following exercises.

EXERCISE 2.6.7. Let $d \geq 2$ and $\Omega = \{-\pi/2 < x^d < \pi/2\}$. Give an example of a nontrivial harmonic (unbounded) function u which is continuous in $\bar{\Omega}$ and $u = 0$ on $\partial\Omega$.

EXERCISE 2.6.8. Let $d = 2$, $\varepsilon > 0$ and $\Omega \subset \{(x, y) : -\pi/2 + \varepsilon < y < \pi/2 - \varepsilon\}$. Let u be a harmonic function in Ω such that $u = 0$ on $\partial\Omega$. Prove that if Ω is unbounded and

$$\limsup_{(x,y)\in\Omega, |x|\to\infty} |u(x, y)|e^{-|x|} = 0,$$

then $u \equiv 0$ in Ω.

EXERCISE 2.6.9. Prove that if $\Omega = \{-\pi/4 < x^d < \pi/4\}$ and a bounded harmonic function $u \in C^2_{loc}(\Omega) \cap C(\bar{\Omega})$, then (2.6.2) holds true.

EXERCISE 2.6.10. Prove that if Ω is a half space and a bounded harmonic function $u \in C^2_{loc}(\Omega) \cap C(\bar{\Omega})$, then (2.6.2) holds true.

EXERCISE 2.6.11. Let $d \geq 2$, $\Omega = \{x \in \mathbb{R}^d : x^d > 0\}$. Assume that $u \in C^2_{loc}(\Omega) \cap C(\bar{\Omega})$ and u is bounded and harmonic in Ω. Finally, take a number $\delta \in (0, 1)$ and assume that for any $x, y \in \partial\Omega$ we have $|u(x) - u(y)| \leq |x - y|^\delta$. By using the maximum principle and Exercise 2.1.4 prove that for a constant $N = N(d, \delta)$ the inequality $|u(x) - u(y)| \leq N|x - y|^\delta$ holds for all $x, y \in \Omega$.

2.7. Poisson's equation in a ball

At first sight the maximum principle does not look like a very deep result. Even more surprising is that if one is only interested in the results below, in these lectures, which are related to the second–order elliptic equations with real coefficients, one might as well start just from the previous section. Nothing before is needed.

In this section we will see how to prove the solvability of Poisson's equation in a ball, first on the basis of the previous results, and then on the basis of the maximum principle alone.

Let $\mathcal{P}_n$ be the set of all polynomials of x of order less than or equal to n. This is a finite–dimensional vector space. Define the following operator on $\mathcal{P}_n$:

$$Tu = \Delta[(1 - |x|^2)u].$$

It is easy to check that $T : \mathcal{P}_n \to \mathcal{P}_n$. Furthermore, if $Tu = 0$, then by the maximum principle $|(1 - |x|^2)u(x)|$ for $|x| \leq 1$ is less than the maximum of this function on $|x| = 1$. The latter is zero, which implies that $u = 0$. From linear algebra one infers that for any $p \in \mathcal{P}_n$ there exists a unique $u \in \mathcal{P}_n$ such that $Tu = p$. The function $(1 - |x|^2)u$ solves the equation $\Delta v = p$ and equals zero on ∂B_1. If we want to solve the same equation with a *polynomial* boundary data g, we can just consider the difference $v - g$ which should satisfy $\Delta w = p - \Delta g$. In this way we arrive at the following theorem.

THEOREM 2.7.1. *Let f, g be some polynomials, then there exists a unique (polynomial) function $u \in C^2(\bar{B}_1)$ such that*

$$\Delta u = f \quad in \quad B_1, \quad u = g \quad on \quad \partial B_1.$$

REMARK 2.7.2. For $f = 0$ the solution u is given by the Poisson integral formula (2.4.2). In particular, the right–hand side of (2.4.2) tends to $g(y)$ if $B_R \ni x \to y \in \partial B_R$ when g is a polynomial. Also, if $g \equiv 1$, $u \equiv 1$, so that

$$\int_{|y|=R} H(x, y) g(y) \, dS_y = 1.$$

Hence

$$|\int_{|y|=R} H(x, y) g_1(y) \, dS_y - \int_{|y|=R} H(x, y) g_2(y) \, dS_y| \leq \max_{|y|=R} |g_1(y) - g_2(y)|.$$

By using this and approximating continuous functions by polynomials, one gets that the right–hand side of (2.4.2) tends to $g(y)$ if $B_R \ni x \to y \in \partial B_R$ for any continuous g.

Now we are ready to prove the solvability of the Dirichlet problem for balls. It is to be noticed that in the future we will prove much stronger results.

THEOREM 2.7.3. *Let $f \in C^2(\mathbb{R}^d)$ and $g \in C(\mathbb{R}^d)$. Then there exists a unique solution $u \in C^2_{loc}(B_1) \cap C(\bar{B}_1)$ of the Dirichlet problem*

$$\Delta u = f \quad in \quad B_1, \quad u = g \quad on \quad \partial B_1.$$

Proof. We may assume that f has compact support. Then by Corollary 2.2.2 we have a solution of the equation $\Delta w = f$ in B_1. Therefore, we further reduce our problem to the case in which $f \equiv 0$. In that case we remember that the Poisson integral formula (2.4.2) defines a harmonic function in B_1, which by the previous Remark 2.7.2 satisfies the boundary condition $u = g$. The theorem is proved.

REMARK 2.7.4. In Theorem 2.7.1 we could also consider ellipsoids instead of balls. Actually parabolas and hyperbolas are also good for this method, since they can be represented as null sets of second–order polynomials. The only inconvenience is that in the latter case solutions are unbounded.

We cannot say the same about Theorem 2.7.3, since its proof depends on the Poisson integral formula (2.4.2). Below we show how to prove Theorem 2.7.3 using the maximum principle alone.

After Theorem 2.7.1 the next natural step is to approximate by polynomials f and g continuous functions and try to understand what is happening to the corresponding solutions. It is desirable to show that they converge to a solution. Let us first consider the case of harmonic functions in which $f = 0$. We want to have control of the smoothness of solutions corresponding to different g. We saw how to do this on the basis of the Poisson integral formula in Theorem 2.5.2.

It turns out that there is a very general method of proving estimates like (2.5.1) without using any integral representations of u. This method works for other second–order elliptic *and parabolic* (linear or nonlinear) equations even with variable coefficients. We have in mind the following Bernstein's method which uses only the maximum principle.

Let u be harmonic in B_1 and assume that $u \in C^3(\bar{B}_1)$. Take any $\zeta \in C_0^\infty(\mathbb{R}^d)$ with support in B_1, assume $\zeta(0) = 1$ and consider the function

$$w := \zeta^2 |u_x|^2 + \lambda |u|^2,$$

where $\lambda > 0$ is a constant. We have $\Delta u = 0$, $\Delta u_{x^i} = 0$, and

$$\Delta w = |u_x|^2 \Delta(\zeta^2) + \zeta^2 [2u_{x^i} \Delta u_{x^i} + 2 \sum_{ij} u^2_{x^i x^j}] + 8\zeta \zeta_{x^i} u_{x^j} u_{x^i x^j}$$

$$+ 2\lambda |u_x|^2 + 2\lambda u \Delta u = |u_x|^2 [2\lambda + \Delta(\zeta^2)] + 2\zeta^2 \sum_{ij} u^2_{x^i x^j} + 8[\zeta_{x^i} u_{x^j}] \cdot [\zeta u_{x^i x^j}]$$

$$\geq |u_x|^2 [2\lambda + \Delta(\zeta^2) - 8|\zeta_x|^2]$$

(we have used $2a^2 + 8ab \geq -8b^2$). We see how to take λ so that $\Delta w \geq 0$. Fix such a λ. Then by the maximum principle

$$|u_x|^2(0) \leq \max_{\bar{B}_1} w \leq \max_{\partial B_1} w = \lambda \max_{\partial B_1} |u|^2.$$

This yields (2.5.1) for $R = 1$ with a different constant (which hardly ever plays any role). For other values of R one gets (2.5.1) by using dilations, that is the fact that if $\Delta u = 0$, then for $v(x) = u(\mu x)$ we have $\Delta v = 0$. As in Theorem 2.5.2 this leads us to the following slightly weaker result.

LEMMA 2.7.5. *Let Ω be a domain and $u \in C(\bar{\Omega})$ be an infinitely differentiable harmonic function in Ω. Then for any multi-index α and any $x \in \Omega$ we have*

$$|D^\alpha u(x)| \leq \left[\frac{N|\alpha|}{\text{dist}\,(x, \partial\Omega)} \right]^{|\alpha|} \sup_{\Omega} |u|, \tag{2.7.1}$$

where the constant N depends only on d.

REMARK 2.7.6. This lemma implies Corollaries 2.5.3 and 2.5.4 with only one additional assumption that all harmonic functions involved are supposed to be infinitely differentiable.

REMARK 2.7.7. Even this weaker version of Corollary 2.5.4 allows us to obtain one more proof of the Fundamental Theorem of Algebra. Indeed, if $p(z)$ is a polynomial of degree ≥ 1, then $\text{Re}\, p^{-1}$ is a C^∞–harmonic function on $\mathbb{R}^2$ outside the set where $p(z) = 0$. If this set of roots is empty, then the harmonic function $\text{Re}\, p^{-1}$ is defined everywhere. In addition, $|\text{Re}\, p^{-1}| \leq |p|^{-1}$, which tends to zero as $|z| \to \infty$; hence $\text{Re}\, p^{-1}$ is bounded and therefore constant. This constant should be zero since $\text{Re}\, p^{-1}$ tends to zero as $|z| \to \infty$. The same is true for $\text{Im}\, p^{-1}$, and this leads us to the false conclusion $|p|^{-1} \equiv 0$.

Now we can prove the following version of Theorem 2.7.3.

THEOREM 2.7.8. *Let $g \in C(\mathbb{R}^d)$. Then there exists a unique solution $u \in C^2_{loc}(B_1) \cap C(\bar{B}_1)$ of the Dirichlet problem*

$$\Delta u = 0 \quad in \quad B_1, \quad u = g \quad on \quad \partial B_1.$$

Moreover, u is infinitely differentiable in B_1, so that the assertion of Lemma 2.7.5 is valid for any harmonic $u \in C^2_{loc}(B_1) \cap C(\bar{B}_1)$ and $\Omega = B_1$.

Proof. Uniqueness we know from the maximum principle. To prove existence, take a sequence of polynomials p_n converging to g uniformly on ∂B_1, and let u_n be the corresponding harmonic functions in B_1 with boundary values g_n. By the maximum principle

$$\max_{B_1} |u_n - u_m| \leq \max_{B_1} |g_n - g_m|,$$

which implies that there is a function u to which u_n converge uniformly in $\bar{B}_1$. By Lemma 2.7.5 applied to $D^\alpha(u_n - u_m)$ the same convergence holds for all derivatives inside any compact subset of B_1. Since $\Delta u_n = 0$, u is harmonic in B_1. Obviously $u \in C(\bar{B}_1)$, u is infinitely differentiable in B_1 and $u = g$ on ∂B_1. The theorem is proved.

We leave to the reader to prove Theorem 2.7.3 for $f \neq 0$ by doing the following exercise.

EXERCISE 2.7.9. Let $u \in C^4(\bar{B}_1)$ and denote $f = \Delta u$. Prove that for any $r \in (0,1)$ there is a constant $N = N(r,d)$ such that in B_r

$$|u| + |u_x| + |u_{xx}| \leq N(\sup_{B_1} |f| + \sup_{B_1} |f_x| + \sup_{B_1} |f_{xx}| + \max_{\partial B_1} |u|).$$

EXERCISE 2.7.10. By using Theorem 2.7.8 and (2.7.1) prove that harmonic functions in Ω are real-analytic in Ω. In particular, if u is harmonic in a connected domain Ω and $u \equiv 0$ in an open subset of Ω, then $u \equiv 0$ in Ω.

EXERCISE 2.7.11 (removable singularities). Let $d \geq 2$ and $u \in C^2_{loc}(B_1 \setminus \{0\})$ be a *bounded* harmonic function in $B_1 \setminus \{0\}$. Prove that $\lim_{x\to 0} u(x) =: a$ exists, and if we define $u(0) = a$, then the function u is harmonic in B_1.

2.8. Other second–order elliptic operators with constant coefficients

Given an operator

$$Lu(x) = a^{ij}(x)u_{x^i x^j}(x) + b^i(x)u_{x^i}(x) + c(x)u(x),$$

we call it second–order elliptic (possibly degenerate) if the matrix $a(x) = (a^{ij}(x))$ is symmetric and nonnegative for any $x \in \mathbb{R}^d$. If there is a constant $\kappa > 0$ such that $a^{ij}(x)\xi^i\xi^j \geq \kappa|\xi|^2$ for all $x, \xi \in \mathbb{R}^d$, we say that L is uniformly elliptic or uniformly nondegenerate. The constant κ is called a constant of ellipticity of L. In this section we assume that a, b, c are independent of x and $c < 0$.

REMARK 2.8.1. The matrix a can be represented as $(\sum_{k=1}^d \lambda^k p^{ki} p^{kj})$, where (p^{ij}) is an orthogonal matrix. It follows that if we introduce new coordinates in $\mathbb{R}^d$ and make the change of variables

$$y^k = p^{ki}x^i, \quad v(y) = u(x),$$

then the equation $Lu(x) = f(x)$ becomes

$$\lambda^i v_{y^i y^i}(y) + (b^k p^{ik}) v_{y^i}(y) + cv(y) = f(x).$$

If, in addition, the matrix a is nondegenerate so that the eigenvalues $\lambda^i > 0$, then one can further transform the equation by introducing $z^i = (\lambda^i)^{-1/2} y^i$, $i = 1, ..., d$ (with no summation in i) and obtaining for $w(z) = v(y) = u(x)$,

$$\Delta w(z) + \sum_{k,i} b^k p^{ik} (\lambda^i)^{-1/2} w_{z^i}(z) + cw(z) = f(x).$$

In particular, investigation of the equation $Lu = f$ with $b = 0, c = 0$ and nondegenerate a can be reduced to investigation of $\Delta w = f$ by a linear change of coordinates.

One of the good features of equations with constant coefficients and with $c < 0$ in the whole space is that one can always find solutions explicitly. We have already met such a situation for nondegenerate equations in the previous chapter. Here we can allow a to degenerate and even vanish. It is worth noting that Theorem 2.8.2 can be considered as a particular case of probabilistic representations of solutions (cf. [**9**]). For $t > 0$ and $x \in \mathbb{R}^d$ we define

$$G(t, x) = \frac{1}{(4\pi t)^{d/2}} \exp(-\frac{1}{4t}|x|^2). \tag{2.8.1}$$

We will use the fact that the function $G(t, x)$ is infinitely differentiable with respect to (t, x) and

$$\Delta G(t, x) - \frac{\partial}{\partial t} G(t, x) = 0, \qquad \int_{\mathbb{R}^d} G(t, x)\, dx = 1. \tag{2.8.2}$$

In the following theorem by σ we mean a nonnegative symmetric square root of the matrix a. If one takes p from Remark 2.8.1, then one has $\sigma^{ij} = \sum_{k=1}^d \sqrt{\lambda^k}\, p^{ki} p^{kj}$.

THEOREM 2.8.2. *Assume that $c < 0$ and for bounded f given on $\mathbb{R}^d$ define*

$$\mathcal{R}f(x) = \int_0^\infty \int_{\mathbb{R}^d} f(x + \sigma y + bt) G(t, y) e^{ct}\, dy\, dt. \tag{2.8.3}$$

(i) If $u \in C_b^2(\mathbb{R}^d)$, then in $\mathbb{R}^d$ we have $u = -\mathcal{R}(Lu)$.

(ii) If $f \in C_b^n(\mathbb{R}^d)$ and $u := -\mathcal{R}f$, then $u \in C_b^n(\mathbb{R}^d)$; and if $n \geq 2$, then $Lu = f$ in $\mathbb{R}^d$.

Proof. Observe that the condition $c < 0$ and the second relation in (2.8.2) guarantee that the integrals in (2.8.3) converge.

(i) For $\varepsilon > 0$ define $\mathcal{R}_\varepsilon f$ by taking the integral in (2.8.3) with respect to t from ε rather than 0. Clearly $\mathcal{R}_\varepsilon f \to \mathcal{R}f$ as $\varepsilon \downarrow 0$. Further, fix $x \in \mathbb{R}^d$ and let $v(t, y) = u(x + \sigma y + bt)$. Simple computations show that $(Lu)(x + \sigma y + bt) = \Delta_y v(t, y) + v_t(t, y) + cv(t, y)$. Therefore, from (2.8.2) integrating by parts we obtain

$$\mathcal{R}_\varepsilon(Lu)(x) = \int_\varepsilon^\infty \int_{\mathbb{R}^d} G(t,y)e^{ct}\Delta_y v(t,y)\,dy\,dt$$

$$+ \int_{\mathbb{R}^d}\int_\varepsilon^\infty G(t,y)(e^{ct}v(t,y))_t\,dt = -e^{c\varepsilon}\int_{\mathbb{R}^d} G(\varepsilon,y)v(\varepsilon,y)\,dy$$

$$= -e^{c\varepsilon}\int_{\mathbb{R}^d} G(\varepsilon,y)u(x+\sigma y+b\varepsilon)\,dy = -e^{c\varepsilon}\int_{\mathbb{R}^d} G(1,y)u(x+\sigma\sqrt{\varepsilon}\,y+b\varepsilon)\,dy,$$

where the last equality follows after a change of variables. By the dominated convergence theorem this implies that $\mathcal{R}_\varepsilon(Lu)(x) \to -u(x)$ as $\varepsilon \downarrow 0$, which proves (i).

(ii) By rules for differentiation of integrals we easily get that $u \in C_b^n(\mathbb{R}^d)$ and $Lu = -\mathcal{R}(Lf)$ if $n \geq 2$. Assertion (i) says that the last expression is f. The theorem is proved.

Interestingly enough Theorem 2.8.2 applies to parabolic equations in the whole space. We mean the equation

$$Lu(s,x) - u_s(s,x) = f(s,x) \quad s \in \mathbb{R}, x \in \mathbb{R}^d. \tag{2.8.4}$$

Actually the operator $L - \partial/\partial s$ is of the same type as L only in the space of one dimension more and with a special notation for one of the coordinates. Taking this into account, we write down $\mathcal{R}f(s,x)$ for this situation as

$$\int_0^\infty \int_{(r,y)\in\mathbb{R}^{d+1}} f(s-t, x+\sigma y+bt)\frac{1}{(4\pi t)^{(d+1)/2}}\exp[-\frac{1}{4t}(r^2+|y|^2)+ct]\,drdy\,dt.$$

One can simplify this expression by integrating out the variable r by using (2.8.2) with $d = 1$. Then we see that

$$\mathcal{R}f(s,x) = \int_0^\infty \int_{\mathbb{R}^d} f(s-t, x+\sigma y+bt)G(t,y)e^{ct}\,dy\,dt, \tag{2.8.5}$$

and from Theorem 2.8.2 we obtain the following result.

THEOREM 2.8.3. *Assume that $c < 0$ and for bounded $f(s,x)$ given on $\mathbb{R}^{d+1}$ define $\mathcal{R}f(s,x)$ by (2.8.5).*

(i) If $u \in C_b^2(\mathbb{R}^{d+1})$ and f is defined by (2.8.4), then in $\mathbb{R}^{d+1}$ we have $u = -\mathcal{R}f$.

(ii) If $f \in C_b^n(\mathbb{R}^{d+1})$ and $u := -\mathcal{R}f$, then $u \in C_b^n(\mathbb{R}^{d+1})$; and if $n \geq 2$, then u satisfies (2.8.4).

2.9. The maximum principle for second–order equations with variable coefficients

In the future we will need the maximum principle for elliptic operators with variable coefficients. We take the operator L from the previous section but allow to the coefficients of L to depend on x: $a^{ij} = a^{ij}(x), b^i = b^i(x), c = c(x)$. We keep the assumptions that the matrix a is symmetric and nonnegative for any x and $c(x) < 0$. In the following theorem we use the simple fact that given two symmetric $d \times d$ nonnegative matrices A and B, we have $\operatorname{tr} AB \geq 0$. One can prove this fact by making one of the matrices diagonal.

THEOREM 2.9.1 (the maximum principle). *Let Ω be a bounded domain in $\mathbb{R}^d$ and $u \in C^2_{loc}(\Omega) \cap C(\bar{\Omega})$. Assume that $Lu \geq 0$ in Ω and $u \leq 0$ on $\partial\Omega$. Then $u \leq 0$ in Ω.*

Proof. Assume the contrary. Then at some point $x_0 \in \Omega$ we have $u(x_0) = \max_\Omega u(x) > 0$. At this point the matrix u_{xx} is symmetric and nonpositive and $u_x = 0$. Therefore,

$$Lu(x_0) = \operatorname{tr} a(x_0)u_{xx}(x_0) + b(x_0) \cdot u_x(x_0) + c(x_0)u(x_0) \le c(x_0)u(x_0) < 0,$$

which contradicts $Lu \ge 0$, and the theorem is proved.

THEOREM 2.9.2. *Let Ω be a domain in $\mathbb{R}^d$ and u be a bounded and continuous function on $\bar{\Omega}$ and $u = 0$ on $\partial\Omega$ if $\partial\Omega \ne \emptyset$ (that is, if $\Omega \ne \mathbb{R}^d$). Moreover, assume that $u \in C^2_{loc}(\Omega)$. Finally, let $a(x), b(x)$ be bounded and $c(x) \le -\lambda$ where the constant $\lambda > 0$. Then in Ω*

$$u \le \frac{1}{\lambda} \sup_\Omega (Lu)_-, \qquad |u| \le \frac{1}{\lambda} \sup_\Omega |Lu|. \tag{2.9.1}$$

Proof. We need prove only the first inequality in (2.9.1), since the second one follows after considering u and $-u$. Define $M = \sup_\Omega (Lu)_-$ and notice that, as follows from simple inequalities, for all sufficiently small $\varepsilon > 0$, we have $L(\varepsilon|x|^2 + 1) \le 0$. Fix an appropriate ε and take a constant $\delta > 0$. For the function $w := u - \delta(\varepsilon|x|^2 + 1) - \lambda^{-1}M$ we have

$$Lw = Lu - \delta L(\varepsilon|x|^2 + 1) - c\lambda^{-1}M \ge -M - c\lambda^{-1}M \ge 0.$$

Moreover, given δ, for all large enough $R > 0$ the function w is negative on $\Omega \cap \partial B_R$, which implies that it is negative on $\partial(\Omega \cap B_R)$. By Theorem 2.9.1 we get $w \le 0$ in $\Omega \cap B_R$ for any large R, that is, $w \le 0$ in Ω. By letting $\delta \downarrow 0$ we conclude $u \le \lambda^{-1}M$. The theorem is proved.

This theorem implies the following uniqueness result.

COROLLARY 2.9.3. *Under the conditions of Theorem 2.9.2 assume that $Lu = 0$ in Ω. Then $u = 0$ in $\bar{\Omega}$.*

EXERCISE* 2.9.4. Keep the assumptions of Theorem 2.9.2, but instead of assuming that $u = 0$ on $\partial\Omega$ assume that $u \le 0$ on $\partial\Omega$. Also assume that $Lu \ge 0$. Prove that $u \le 0$ in Ω.

EXERCISE 2.9.5. Keep the assumptions of Theorem 2.9.2, but instead of assuming that a, b are bounded assume that $\operatorname{tr} a(x) + x \cdot b(x) \le K(1 + |x|^2)$ for all x, where K is a constant. Prove that in this situation the assertions of the theorem are true as well.

It is interesting that the result of this exercise is essentially sharp. Specifically one can prove that for $d = 1$ the equation

$$u'' + x^3 u' - u = 0 \quad \text{on} \quad (-\infty, \infty)$$

has a nontrivial bounded solution (actually any solution of this equation is bounded). Also the same is true if one replaces x^3 by $|x|^r \operatorname{sign} x$ with any $r > 1$.

EXERCISE 2.9.6. Prove Gauss's law of the arithmetic mean in the following situation. Assume that $c \equiv 0$ and that the operator L is invariant under any orthogonal transformation: $Lu(p\cdot)(x) = (Lu)(px)$ for any orthogonal p and smooth u. Also assume that there is a function $v \in C^2(\bar{B}_1)$ such that $Lv \le -1$ in B_1 and

$v > 0$ on ∂B_1. Let $u \in C^2(\bar{B}_1)$ and $Lu = 0$ in B_1. By using the maximum principle prove that

$$u(0) = \frac{1}{\omega_d} \int_{\partial B_1} u(y)\, dS.$$

2.10. Hints to exercises

2.3.3. To prove that $u(x, y) \to f(x)$ as $y \downarrow 0$, change the variable by letting $t = x + ys$.
2.3.4. Represent the right–hand side of (2.3.3) as

$$\frac{1}{\omega_d} \int_{|y|=1} u(Ry)\, dS,$$

then differentiate with respect to R and use Gauss's divergence theorem.
2.3.5. From Exercise 2.3.4 it follows that one need prove only that u is infinitely differentiable. To prove this, multiply (2.3.4) by a function $\mathbb{R}^{d-1}\psi(R)$ with compact support and integrate with respect to R. Show that for appropriate ψ and x one obtains $cu(x) = \psi(|x|) * u(x)$ where c is a constant.
2.6.4. Consider $u = \pm w - v \sup(f_{\mp})/(2d)$.
2.6.5. Remember Gauss's law of the arithmetic mean.
2.6.7. Try $(\cos x^d) v(x^1)$.
2.6.8. Observe that $u_0(x, y) := (\cos y) \cosh x$ is harmonic in Ω and for any $\gamma > 0$ there exists $R > 0$ such that $u \leq \gamma u_0$ on $\partial(\Omega \cap \{|x| \leq R\})$.
2.6.9. Show that for $\varepsilon > 0$ the function $u(x) - \varepsilon(\cos x^d)(\cosh x^1 + ... + \cosh x^{d-1})$ must take its maximum value in $\bar{\Omega}$ and that it occurs on $\partial\Omega$.
2.6.10. Let $\Omega = \{x^d > 0\}$, and for $R > 0$ let $\Omega_R = \{R > x^d > 0\}$. Define $v(x) = x^d R^{-1} \sup_{x^d = R} u + (1 - x^d R^{-1}) \sup_{x^d = 0} u$, and by using Exercise 2.6.9 show that $u \leq v$ in Ω_R. Then let $R \to \infty$.
2.6.11. Let $w(x, y)$ be the function from Exercise 2.1.4 and for $x \in \Omega$ define

$$v(x) = w(x^1, x^d) + ... + w(x^{d-1}, x^d).$$

Observe that $u(x) - u(0) - w \leq 0$ on $\partial\Omega$ and tends to $-\infty$ as $|x| \to \infty$, so that the set where it is nonnegative is bounded. Conclude that $u(x) - u(0) \leq N|x|^\delta$, then make the origin vary. Finally, for fixed $y \in \Omega$ consider the function $u(x + y) - u(x)$.
2.7.9. Define the function $w = \zeta^2 |u_x|^2 + \lambda |u|^2$ with ζ as before, and prove that for some constants $\lambda > 0$ and $\mu > 0$ one has

$$\Delta w \geq -\mu(\sup_{B_1} |u|^2 + \sup_{B_1} |f|^2 + \sup_{B_1} |f_x|^2)$$

in B_1. From Exercise 2.6.4 derive that $|u_x|$ in B_r with $r < 1$ is controlled by $\sup_{B_1} |f| + \sup_{B_1} |f_x|$ and $\max_{\partial B_1} |u|$. Dilations allow one to get a similar result in B_R instead of B_1. Finally, apply this argument to the equation $\Delta u_{x^i} = f_{x^i}$.
2.7.10. Use Taylor's formula and make the number of terms go to infinity.
2.7.11. Solve the equation $\Delta v = 0$ in $B_{1/2}$ with boundary data $v = u$ on $\partial B_{1/2}$. Then define $w_{\pm} = u - v \pm \varepsilon\psi$, where $\varepsilon > 0$ and ψ is the (negative) fundamental solution of $\Delta\psi = \delta_0$, and observe that $\mp w_{\pm}$ goes to infinity as x goes to zero. After this apply the maximum principle to show that $\mp w \geq 0$ in $B_1 \setminus \{0\}$ for any ε. Conclude that $u = v$ in $B_1 \setminus \{0\}$.

2.9.5. Define $v(x) = 1 + |x|^2$ and check that $(L - 2K)v \leq 0$. Then by repeating the proof of Theorem 2.9.2 with $w = u - \delta v - (\lambda + 2K)^{-1}M$, where $M = \sup_\Omega(Lu - 2Ku)_-$, show that

$$\sup_\Omega u_+ \leq \frac{1}{\lambda + 2K} \sup_\Omega (Lu - 2Ku)_- \leq \frac{1}{\lambda + 2K} \sup_\Omega (Lu)_- + \frac{2K}{\lambda + 2K} \sup_\Omega u_+.$$

2.9.6. First, by introducing the operator $L'w := L(vw)$ obtain the following uniqueness result: if a function $\psi \in C^2(\bar{B}_1)$ satisfies $L\psi = 0$ in B_1 and $\psi = 0$ on ∂B_1, then $\psi = 0$ in B_1. Then define

$$\psi(x) = \int u(px)\, \pi(dp),$$

where the integral is taken over the orthogonal group and π is the Haar measure. Observe that $L\psi = 0$ in B_1 and that ψ depends only on $|x|$, so that ψ is constant on ∂B_1. Finally, show that for $|x| = 1$

$$\psi(x) = \frac{1}{\omega_d} \int_{|y|=1} u(y)\, dS.$$

CHAPTER 3

Solvability of Elliptic Equations with Constant Coefficients in the Hölder Spaces

Given an mth order elliptic operator L, our general goal is to find conditions as wide as possible on f which guarantee that the elliptic equation $Lu = f$ is solvable. Theorem 1.5.2 says that if $f \in C^m(\mathbb{R}^d)$, then there is a unique function $u \in C^m(\mathbb{R}^d)$ which solves the equation. On the other hand, $L : C^m(\mathbb{R}^d) \to C(\mathbb{R}^d)$, and the question arises if for any $f \in C(\mathbb{R}^d)$ we can find a solution in $C^m(\mathbb{R}^d)$. This is a question of bijectivity of the operator $L : C^m(\mathbb{R}^d) \to C(\mathbb{R}^d)$. Interestingly enough the operator L is *not* bijective in spaces $C^m(\mathbb{R}^d)$. Even in the case of Laplace's operator and the equation $(\Delta - 1)u = f$ it is not true that for any $f \in C(\mathbb{R}^d)$ the solution $u \in C^2(\mathbb{R}^d)$. It turns out that in order to get the bijectivity we have to consider the so-called Hölder spaces $C^{m+\delta}(\mathbb{R}^d)$, and in these spaces we can even solve equations with variable coefficients.

3.1. The Hölder spaces

Let Ω be a domain in $\mathbb{R}^d$. Recall that for $k = 0, 1, 2, \ldots$, we denote $C^k_{loc}(\Omega)$ the set of all functions $u = u(x)$ whose derivatives $D^\alpha u$ for $|\alpha| \le k$ are continuous in Ω. We set

$$|u|_{0;\Omega} = [u]_{0;\Omega} = \sup_{\Omega} |u|, \quad [u]_{k;\Omega} = \max_{|\alpha|=k} |D^\alpha u|_{0;\Omega}. \tag{3.1.1}$$

DEFINITION 3.1.1. For $k = 0, 1, 2, \ldots.$ the space $C^k(\Omega)$ is the Banach space of all functions $u \in C^k_{loc}(\Omega)$ for which the following norm

$$|u|_{k;\Omega} = \sum_{j=0}^{k} [u]_{j;\Omega} \tag{3.1.2}$$

is finite. If $0 < \delta < 1$, we call u *Hölder continuous with exponent δ in Ω* if the seminorm

$$[u]_{\delta;\Omega} = \sup_{\substack{x,y\in\Omega \\ x\neq y}} \frac{|u(x) - u(y)|}{|x - y|^\delta} \tag{3.1.3}$$

is finite. This seminorm is called *Hölder's constant* of u of order δ. If the right-hand side of (3.1.3) is finite for $\delta = 1$, the function u is called *Lipschitz continuous in Ω*. We set

$$[u]_{k+\delta;\Omega} = \max_{|\alpha|=k} [D^\alpha u]_{\delta;\Omega}. \tag{3.1.4}$$

DEFINITION 3.1.2. For $0 < \delta < 1$ and $k = 0, 1, 2, \ldots$ the Hölder space $C^{k+\delta}(\Omega)$ is the Banach space of all functions $u \in C^k(\Omega)$ for which the norm

$$|u|_{k+\delta;\Omega} = |u|_{k;\Omega} + [u]_{k+\delta;\Omega} \tag{3.1.5}$$

is finite.

For simplicity we drop the subscript Ω if $\Omega = \mathbb{R}^d$.

REMARK 3.1.3. Probably we need to show that $C^{k+\delta}(\Omega)$ is a Banach space indeed.

As always it suffices to check the completeness of $C^{k+\delta}(\Omega)$. Take a Cauchy sequence u_n in $C^{k+\delta}(\Omega)$. These functions are equibounded and equicontinuous in any compact subdomain in Ω. Therefore, there exists a subsequence $u_{n(i)}$ which converges to a function u uniformly in any compact subdomain of Ω. Obviously u is bounded and continuous in Ω. The same argument can be repeated for any derivative of u_n of order up to and including k. Moreover, the corresponding subsequences can be chosen to be the same, and then from calculus we obtain that u has k derivatives which are bounded and continuous in Ω. Next, for any multi–index α with $|\alpha| = k$ and any $x, y \in \Omega$ we have

$$\begin{aligned}
&|[D^\alpha u(x) - D^\alpha u(y)] - [D^\alpha u_n(x) - D^\alpha u_n(y)]| \\
&\qquad \le \limsup_{m\to\infty} |[D^\alpha u_m(x) - D^\alpha u_m(y)] - [D^\alpha u_n(x) - D^\alpha u_n(y)]| \\
&\qquad\qquad \le |x-y|^\delta \limsup_{m\to\infty} [D^\alpha u_m - D^\alpha u_n]_{\delta;\Omega}.
\end{aligned}$$

In the same way we can consider $D^\alpha u(x) - D^\alpha u_n(x)$ for $|\alpha| \le k$. It follows that

$$|u - u_n|_{k+\delta;\Omega} \le \limsup_{m\to\infty} |u_m - u_n|_{k+\delta;\Omega},$$

and since the last expression goes to zero as $n \to \infty$, we conclude that the Cauchy sequence u_n converges in the norm of $C^{k+\delta}(\Omega)$ to u.

Observe that the space $C^r(\Omega)$ is defined for all real $r \ge 0$, but for integer $r \ge 1$ it is *not* the space of functions which have $r - 1$ Lipschitz continuous derivatives.

We will also use similar notation for closed domains $\overline{\Omega}$. It is important to notice that if $u \in C^\delta(\Omega)$, then u is uniformly continuous in Ω, and therefore the Cauchy criterion easily implies that u can be uniquely extended to $\partial\Omega$ to give a continuous function in $\bar{\Omega}$. This allows us to speak about values on $\partial\Omega$ of a function $u \in C^\delta(\Omega)$ and allows us to write $C^\delta(\Omega) = C^\delta(\bar{\Omega})$.

From the elementary inequality

$$|u(x)v(x) - u(y)v(y)| \le |u(x)| \cdot |v(x) - v(y)| + |v(y)| \cdot |u(x) - u(y)|$$

and (3.1.1), (3.1.3) we have

$$[uv]_{\delta;\Omega} \le |u|_{0;\Omega} \cdot [v]_{\delta;\Omega} + |v|_{0;\Omega} \cdot [u]_{\delta;\Omega} \tag{3.1.6}$$

for $u, v \in C^\delta(\Omega)$.

LEMMA 3.1.4. *Let $u \in C^k(B_r)$, $|\alpha| = k$. Then there exists $y \in B_r$ such that*

$$|D^\alpha u(y)| \le \frac{(2k)^k}{r^k} |u|_{0;B_r}. \tag{3.1.7}$$

Proof. We take $h = r/k$ and consider the finite-difference operator

$$\delta_h^\alpha u(x) = \delta_{h,1}^{\alpha_1} \delta_{h,2}^{\alpha_2} \cdots \delta_{h,d}^{\alpha_d} u(x), \tag{3.1.8}$$

where

$$\delta_{h,j} u(x) = \frac{1}{h}[u(x + he_j) - u(x)], \quad e_j \text{ is the } j\text{th coordinate vector in } R^d.$$

By the mean value theorem we have $\delta_{h,j} u(x) = u_{x^j}(y)$, where $|x - y| \le h$. Therefore,

$$\delta_{h,i}\delta_{h,j} u(x) = (\delta_{h,j} u)_{x^i}(y_1) = \delta_{h,j}(u_{x^i})(y_1) = u_{x^j x^j}(y_2)$$

where $|x - y_2| \le 2h$. In general, $\delta_h^\alpha u(0) = D^\alpha u(y)$ for some $y \in B_r$, and (3.1.7) follows from the elementary estimate $|\delta_{h,j} u(x)| \le \frac{2}{h}|u|_{0;B_\rho}$ if $|x| + h \le \rho$. The lemma is proved.

EXERCISE* 3.1.5. Let $\Omega \subset \Omega'$ and let the domain Ω' be convex. Prove that if $u \in C^{k+\delta}(\Omega')$ and $u = 0$ on $\Omega' \setminus \Omega$, then $[u]_{k+\delta;\Omega'} = [u]_{k+\delta;\Omega}$.

EXERCISE* 3.1.6. Let $\Omega \subset \Omega'$ and $\rho > 0$ and denote by Ω_ρ the ρ-neighborhood of Ω. Prove that $|u|_{k+\delta;\Omega'} \le (1 + \rho^{-\delta})|u|_{k+\delta;\Omega_\rho \cap \Omega'}$ if $u \in C^{k+\delta}(\Omega')$ and $u = 0$ on $\Omega' \setminus \Omega$.

EXERCISE 3.1.7. Let $u(x)$ be defined in a connected domain $\Omega \subset R^n$, and let

$$[u]_{\delta;\Omega} = \sup_{x,y \in \Omega} \frac{|u(x) - u(y)|}{|x - y|^\delta} < \infty$$

for some $\delta > 1$. Show that $u = \text{const}$ in Ω.

3.2. Interpolation inequalities

The following theorem belongs to a series of results on the so–called *interpolation inequalities* (see [**6**], Sec. 6.8). If Ω is an open cone, then by $\theta = \theta(\Omega)$ we denote the smallest distance from the vertex to centers of unit balls which lie inside Ω.

THEOREM 3.2.1. *Let $0 \le s \le r$, and let Ω be an open convex cone and $u \in C^r(\Omega)$. Then for a constant $N = N(d, r, \theta(\Omega))$ and any $\varepsilon > 0$*

$$[u]_{s;\Omega} \le N\varepsilon^{r-s}[u]_{r;\Omega} + N\varepsilon^{-s}|u|_{0;\Omega}. \tag{3.2.1}$$

Proof. We may assume that the vertex of Ω is at the origin. Observe that for any constant $c > 0$ and for $v(x) = u(cx)$ we have

$$[v]_{s;\Omega} = c^s[u]_{s;\Omega}.$$

This easily implies that in order to prove (3.2.1) with arbitrary $\varepsilon > 0$ we need only prove that for $0 \le s \le r$

$$[u]_{s;\Omega} \le N(d,r,\theta)[u]_{r;\Omega} + N(d,r,\theta)|u|_{0;\Omega}. \tag{3.2.2}$$

Represent s, r as $s = j+\gamma$ and $r = k+\delta$, where $j, k = 0, 1, 2, ...$ and $0 \le \gamma, \delta \le 1$. To prove (3.2.2) at first we assume that $k = j$ and prove that with a constant $N = N(d)$

$$[u]_{j+\gamma;\Omega} \le N[u]_{j+\delta;\Omega} + N[u]_{j;\Omega}. \tag{3.2.3}$$

Obviously we may consider only the case $\gamma > 0$.

1. Case $\delta = 1$. Since Ω is convex, by applying the mean value theorem, for $|\alpha| = j$ we get

$$|D^\alpha u(x) - D^\alpha u(y)| \le N|x-y| \cdot [u]_{j+1;\Omega}$$

with a constant N depending only on d. Hence, if $|x-y| \le 1$, then

$$|D^\alpha u(x) - D^\alpha u(y)| \le N|x-y|^\gamma \cdot [u]_{j+\delta;\Omega},$$

and if $|x-y| \ge 1$, then

$$|D^\alpha u(x) - D^\alpha u(y)| \le 2[u]_{j,\Omega} \le 2|x-y|^\gamma \cdot [u]_{j,\Omega}.$$

This clearly implies (3.2.3).

2. Case $\delta < 1$. This case is similar to the previous one since ($\gamma \le \delta$)

$$|D^\alpha u(x) - D^\alpha u(y)| \le |x-y|^\gamma \frac{|D^\alpha u(x) - D^\alpha u(y)|}{|x-y|^\delta} |x-y|^{\delta-\gamma}.$$

Next we prove that if $j \ge 1$, then

$$[u]_{j;\Omega} \le N[u]_{j+\delta;\Omega} + N|u|_{0;\Omega}. \tag{3.2.4}$$

Here we need consider only $\delta > 0$. Given $x \in \Omega$ and $|\alpha| = j$, for any point $x_0 \in \Omega$ with $|x - x_0| = \theta(\Omega)$ we have

$$|D^\alpha u(x)| \le |D^\alpha u(x) - D^\alpha u(x_0)| + |D^\alpha u(x_0)| \le N[u]_{j+\delta,\Omega} + |D^\alpha u(x_0)|.$$

Next use Lemma 3.1.4 after choosing x_0 so that $\text{dist}\,(x_0, \partial\Omega) \ge 1$ and $y \in B_1(x_0)$ so that $|D^\alpha u(y)| \le N|u|_{0;\Omega}$. Then

$$|D^\alpha u(x)| \le N[u]_{j+\delta,\Omega} + |D^\alpha u(x_0)| \le N[u]_{j+\delta,\Omega}$$
$$+ |D^\alpha u(x_0) - D^\alpha u(y)| + N|u|_{0;\Omega} \le N[u]_{j+\delta,\Omega} + N|u|_{0;\Omega}.$$

This proves (3.2.4).

Now if $j = k$, then (3.2.2) follows from (3.2.3) and (3.2.4). If $j < k$, then by (3.2.3) and (3.2.4)

$$[u]_{j+\gamma;\Omega} \le N[u]_{j+1;\Omega} + N|u|_{0;\Omega} \le ... \le N[u]_{k;\Omega} + N|u|_{0;\Omega},$$

and we can again apply (3.2.4). The theorem is proved.

EXERCISE* 3.2.2. Let Ω be an open convex cone. Prove that if we are given a sequence of $u_n \in C^{k+\delta}(\Omega)$ such that $[u_n]_{k+\delta;\Omega} + |u_n|_{0;\Omega}$ is bounded, then there is a subsequence $\{v_n\}$ of $\{u_n\}$ converging to a function u in the sense of $C^{j+\gamma}(\Gamma)$ whenever $j+\gamma < k+\delta$ and compact $\Gamma \subset \bar{\Omega}$. Also prove that $u \in C^{k+\delta}(\Omega)$ and $|u|_{k+\delta;\Omega} \leq \liminf |v_n|_{k+\delta;\Omega}$.

EXERCISE* 3.2.3. Let $\zeta \in C_0^\infty(\mathbb{R}^d), \zeta \geq 0$ and $\int \zeta\,dx = 1$. For $\varepsilon > 0$ define $\zeta_\varepsilon(x) = \varepsilon^{-d}\zeta(x\varepsilon^{-1})$ and let $u^\varepsilon := \zeta_\varepsilon * u$. Prove that

(i) for any bounded measurable u we have $u^\varepsilon \in C_b^\infty(\mathbb{R}^d)$;

(ii) if $u \in C^{k+\delta}(\mathbb{R}^d)$, then

$$|u^\varepsilon|_{k+\delta} \leq |u|_{k+\delta}, \quad |u|_{k+\delta} = \lim_{\varepsilon\downarrow 0} |u^\varepsilon|_{k+\delta}.$$

EXERCISE* 3.2.4. Let $\zeta \in C_0^\infty(\mathbb{R}^d)$, $\zeta(0) = 1$ and $u \in C^{k+\delta}(\mathbb{R}^d)$. Prove that $[u]_{k+\delta} \leq \liminf_{R\to\infty}[u\zeta_R]_{k+\delta}$, where $\zeta_R(x) = \zeta(x/R)$.

EXERCISE 3.2.5. Prove that for $0 < \delta < 1$ the space $C^\delta([0,1])$ is *not* separable.

EXERCISE* 3.2.6. Under the conditions of Theorem 3.2.1 prove the following *multiplicative inequality*: $[u]_{s;\Omega} \leq N(d,r,\theta(\Omega))[u]_{r;\Omega}^{s/r}|u|_{0;\Omega}^{1-s/r}$. In particular, $|u_x|_{0;\Omega}^2 \leq N(d,\theta(\Omega))[u]_{2;\Omega}|u|_{0;\Omega}$. It is easy to memorize these inequalities if one takes into account that the inequalities are invariant under multiplication of u by a constant and also under dilations.

EXERCISE 3.2.7 (first step toward quasilinear equations). Prove that if $F(x,\xi)$ is a function of $x,\xi \in \mathbb{R}^d$ and for some constants $n,k \geq 0$ we have $[F(\cdot,\xi)]_\delta \leq |\xi|^n$ and $|F_\xi| \leq |\xi|^k$, then for any real-valued $u \in C^2(\mathbb{R}^d)$ it holds that

$$[F(\cdot,u_x(\cdot))]_\delta \leq N(d)\{([u]_2|u|_0)^{n/2} + ([u]_2|u|_0)^{k/2}[u]_2^\eta|u|_0^{1-\eta}\},$$

where $\eta = (1+\delta)/2$.

EXERCISE 3.2.8. Let $F(x,\xi)$ be a function of $x \in \mathbb{R}^d$ and $\xi \in \mathbb{R}^{d_1}$. Assume that for some constants $\delta_1,\delta_2 \in (0,1]$ we have

$$|F(x,\xi) - F(y,\xi)| \leq |x-y|^{\delta_1\delta_2}, \quad |F(x,\xi) - F(x,\eta)| \leq |\xi-\eta|^{\delta_2}$$

for any $x,y \in \mathbb{R}^d$ and $\xi,\eta \in \mathbb{R}^{d_1}$. For simplicity of formulation also assume that $\delta_1\delta_2 < 1$. Prove that for any $\mathbb{R}^{d_1}$-valued function $u \in C^{\delta_1}(\mathbb{R}^d)$ we have $F(\cdot,u(\cdot)) \in C^{\delta_1\delta_2}(\mathbb{R}^d)$. Surprisingly enough, generally speaking, the operator $F: u \to F(\cdot,u(\cdot))$ is *not* continuous as an operator from $C^{\delta_1}(\mathbb{R}^d)$ to $C^{\delta_1\delta_2}(\mathbb{R}^d)$ if $\delta_1,\delta_2 \in (0,1)$.

EXERCISE 3.2.9 (cf. Exercise 3.2.6). Prove the following pointwise interpolation inequality: if $u \in C^2(\mathbb{R}^d)$ *and* $u \geq 0$, then

$$|u_x(x)|^2 \leq N(d)u(x)[u]_2.$$

Conclude that if $u \in C^2(\mathbb{R}^d)$ and $u \geq 0$, then $\sqrt{u}$ is Lipschitz continuous.

EXERCISE 3.2.10 (toward Sobolev's embedding theorem). Define u^ε as in Exercise 3.2.3 and take $p \geq 1$. Upon observing that $D^\alpha u^\varepsilon = (D^\alpha \zeta_\varepsilon) * u$, by using Hölder's inequality and interpolating between integers prove that for any $r \geq 0$

$$[u^\varepsilon]_r \leq N\varepsilon^{-r-d/p}||u||_{L_p},$$

where $N = N(d, r, p)$ and

$$||u||_{L_p} := \left(\int_{\mathbb{R}^d} |u(x)|^p \, dx\right)^{1/p}.$$

EXERCISE 3.2.11 (more interpolation inequalities). Observe that $u_{x^i}(x) = [u_{x^i}(x) - \delta_{h,i}u(x)] + \delta_{h,i}u(x) \leq h[u]_2 + h^{\delta-1}[u]_\delta$ for any $h > 0$, and prove that $[u]_1 \leq N[u]_2^{1-1/(2-\delta)}[u]_\delta^{1/(2-\delta)}$ for $u \in C^2(\mathbb{R}^d)$. Apply this and Exercise 3.2.6 to prove that for any integer $r \geq 1$ and $u \in C^r(\mathbb{R}^d)$ one has

$$[u]_1 \leq N[u]_r^{(1-\delta)/(r-\delta)}[u]_\delta^{(r-1)/(r-\delta)}.$$

The most general multiplicative inequalities can be found in Exercise 3.3.7.

3.3. Equivalent norms in the Hölder spaces

Hölder continuous functions can be characterized by the rate of their approximation by polynomials. For $k = 1, 2, ...$ denote by $\mathcal{P}_k$ the set of all polynomials of $x \in \mathbb{R}^d$ of degree at most k. We will also use the following notation for Taylor's polynomial of degree k of a function u at a point $y \in \mathbb{R}^d$

$$T_y^k u(x) := \sum_{|\alpha| \leq k} \frac{1}{\alpha!}(x-y)^\alpha D^\alpha u(y).$$

Finally, for balls $B_\rho(x)$ and a function u let

$$E_k[u; B_\rho(x)] := \inf_{p \in \mathcal{P}_k} |u - p|_{0;B_\rho(x)}, \quad [u]'_{k+\delta} := \sup_{\rho > 0, x \in \mathbb{R}^d} \rho^{-k-\delta} E_k[u; B_\rho(x)].$$

THEOREM 3.3.1. *Let $\delta \in (0, 1]$. Then the seminorms $[\cdot]_{k+\delta}$ and $[\cdot]'_{k+\delta}$ are equivalent in the sense that there is a constant $N = N(d, k, \delta)$ such that for any $u \in C^{k+\delta}(\mathbb{R}^d)$ we have*

$$[u]'_{k+\delta} \leq N[u]_{k+\delta}, \quad [u]_{k+\delta} \leq N[u]'_{k+\delta}. \tag{3.3.1}$$

Proof. To prove the first inequality it suffices to observe that by virtue of Taylor's formula, given ρ and x, for some points $\xi_y \in B_\rho(x)$,

$$E_k[u; B_\rho(x)] \leq |u - T_x^k u|_{0;B_\rho(x)} = \sup_{y \in B_\rho(x)} |\sum_{|\alpha|=k} \frac{1}{\alpha!}[D^\alpha u(\xi_y) - D^\alpha u(x)](y-x)^\alpha|$$

$$\leq N \sup_{|\alpha|=k} \sup_{y \in B_\rho(x)} |D^\alpha u(\xi_y) - D^\alpha u(x)|\rho^k \leq N[u]_{k+\delta}\rho^{k+\delta},$$

where $\alpha! = \alpha_1! \cdot ... \cdot \alpha_d!$.

On the other hand, take $|\alpha| = k$ and recall that for the operator δ_h^α introduced in (3.1.8) and any $x \in \mathbb{R}^d$ and $u \in C_{loc}^k(\mathbb{R}^d)$ there is a point y such that $|x - y| \leq kh$

and $\delta_h^\alpha u(x) = D^\alpha u(y)$. It follows, in particular, that if $p \in \mathcal{P}_k$, then $\delta_h^\alpha p(x)$ is independent of x. Also for any $p \in \mathcal{P}_k$, $u \in C^{k+\delta}(\mathbb{R}^d)$, $x_1, x_2 \in \mathbb{R}^d$

$$|D^\alpha u(x_1) - D^\alpha u(x_2)| \leq |D^\alpha u(x_1) - \delta_h^\alpha u(x_1)| + |\delta_h^\alpha u(x_1) - \delta_h^\alpha u(x_2)|$$

$$+ |D^\alpha u(x_2) - \delta_h^\alpha u(x_2)| \leq 2[u]_{k+\delta}(kh)^\delta + |\delta_h^\alpha (u-p)(x_1) - \delta_h^\alpha (u-p)(x_2)|$$

$$\leq 2[u]_{k+\delta}(kh)^\delta + |\delta_h^\alpha (u-p)(x_1)| + |\delta_h^\alpha (u-p)(x_2)|$$

$$\leq 2[u]_{k+\delta}(kh)^\delta + Nh^{-k}|u-p|_{0,B_r(x_0)},$$

where $2x_0 = x_1 + x_2$, $r = |x_1 - x_2|/2 + 2kh$. Since p is arbitrary, we get

$$|D^\alpha u(x_1) - D^\alpha u(x_2)| \leq 2[u]_{k+\delta}(kh)^\delta + Nh^{-k}r^{k+\delta}[u]'_{k+\delta}. \tag{3.3.2}$$

Finally take $h = \varepsilon|x_1 - x_2|$ with a fixed constant $\varepsilon > 0$, then in (3.3.2) divide through by $|x_1 - x_2|^\delta$ and take supremum over x_1, x_2 and α on the left. Then

$$[u]_{k+\delta} \leq 2[u]_{k+\delta}(k\varepsilon)^\delta + N[u]'_{k+\delta}.$$

For ε such that $2(k\varepsilon)^\delta = 1/2$ the last inequality gives us the second inequality in (3.3.1), and the theorem is proved.

EXERCISE 3.3.2. Let $\delta \in (0,1]$. Define

$$[u]''_{k+\delta} = \sup_{y\in\mathbb{R}^d} \sup_{\rho>0} \rho^{-k-\delta} \sup_{|x-y|\leq\rho} |u(x) - T_y^k u(x)|$$

and prove that all three seminorms $[\,\cdot\,]_{k+\delta}, [\,\cdot\,]'_{k+\delta}, [\,\cdot\,]''_{k+\delta}$ are equivalent on $C^{k+\delta}(\mathbb{R}^d)$.

EXERCISE 3.3.3. If T is any bounded operator and I is the unit operator, then $T - I = (1/2)\{(T^2 - I) - (T-I)^2\}$. By using this for the shift operators $T_h : u \to u(\cdot + h)$, where $h \in \mathbb{R}^d$ is fixed and the function u is defined on $\mathbb{R}^d$, prove that if $u \in C^\delta(\mathbb{R}^d)$ and $\delta \in (0,1)$, then

$$[u]_\delta = \sup_{|h|>0} |h|^{-\delta}|(T_h - I)u|_0,$$

$$[u]_\delta \leq N(d,\delta) \sup_{|h|>0,x} \frac{|u(x+h) - 2u(x) + u(x-h)|}{|h|^\delta} \leq 2N(d,\delta)[u]_\delta.$$

By the way, bounded functions for which the middle term is finite for $\delta = 1$ form the so–called Zygmund's space $\mathcal{C}^1$.

EXERCISE 3.3.4. One can use other approximations to characterize Hölder continuous functions. Define u^ε as in Exercise 3.2.3 and let

$$[u]''_\delta = \sup_{\varepsilon>0} \varepsilon^{1-\delta}[u^\varepsilon]_1.$$

Prove that for $u \in C^\delta(\mathbb{R}^d)$ and $\delta \in (0,1]$ the seminorms $[u]''_\delta$ and $[u]_\delta$ are equivalent.

EXERCISE 3.3.5 (sharp Sobolev's embedding theorem). Let $d < p < \infty$ and define $\delta = 1 - d/p \in (0,1)$. By combining Exercises 3.3.4 and 3.2.10 prove that for any $u \in C_0^\infty(\mathbb{R}^d)$ we have $[u]_\delta \leq N||u_x||_{L_p}$, where $N = N(d,p)$.

EXERCISE 3.3.6. Take an integer n and let $0 < r \leq n$. Define u^{ε} as in Exercise 3.2.3 and let

$$[u]''_r = \sup_{\varepsilon>0} \varepsilon^{n-r}[u^{\varepsilon}]_n.$$

Prove that for $u \in C^r(\mathbb{R}^d)$ the seminorms $[u]''_r$ and $[u]_r$ are equivalent.

EXERCISE 3.3.7. Observe that the function $[u]''_r$ from Exercise 3.3.6 is obviously convex with respect to r and prove the following general multiplicative inequality: $t > 0$, $0 \leq s \leq r$ and $u \in C^{r+t}(\mathbb{R}^d)$, then $[u]_{s+t} \leq N(d,r,t)[u]^{s/r}_{r+t}[u]^{1-s/r}_t$.

EXERCISE 3.3.8. Take an integer $n \geq 0$ and $\delta \in (0,1]$. Define u^{ε} as in Exercise 3.2.3. Let

$$[u]''_{\delta} = \sup_{\varepsilon>0} \varepsilon^{n}[u^{\varepsilon}]_{n+\delta}.$$

Prove that for $u \in C^{\delta}(\mathbb{R}^d)$ the seminorms $[u]''_{\delta}$ and $[u]_{\delta}$ are equivalent.

3.4. A priori estimates in the whole space for Laplace's operator

One of the basic ideas of the modern theory of partial differential equations consists in using so-called a priori estimates. An a priori estimate is an estimate of a function through the result of application of an operator to this function. It turns out that in almost all situations when an appropriate a priori estimate can be obtained, one also can prove solvability of the corresponding equation. We will be talking about a priori estimates in the Hölder spaces. *Starting with this section everywhere below in these lectures we assume that* $0 < \delta < 1$ *unless a different assumption is stated explicitly.*

The usefulness of a priori estimates is not obvious a priori, so that after Hölder, Lichtenstein, Korn, Hopf and others obtained first a priori estimates in the Hölder spaces for Laplace's operator, it took several more years before people realized their full strength. The estimates which we prove in this section present very particular cases of estimates from Section 3.6 where we consider operators of arbitrary order $m \geq 2$. Therefore, if the reader feels comfortable enough with the general theory, he can skip this section.

THEOREM 3.4.1. *There exists a constant* $N = N(d,\delta)$ *such that for any* $u \in C^{2+\delta}(\mathbb{R}^d)$ *we have*

$$[u]_{2+\delta} \leq N[\Delta u]_{\delta}. \tag{3.4.1}$$

Proof. We follow an idea of Safonov. Denote $f = \Delta u$, take a constant $K \geq 1$ to be chosen later, and take $\rho > 0$. Also let h be a solution of $\Delta h = 0$ in $B_{(K+1)\rho}$ with boundary data $h = u - T^2_0 u$ on $\partial B_{(K+1)\rho}$ (see Sec. 2.7). The function

$$w(x) = u(x) - T^2_0 u(x) - h(x)$$

satisfies $\Delta w(x) = f(x) - f(0)$ in $B_{(K+1)\rho}$ and $w = 0$ on $\partial B_{(K+1)\rho}$. By Exercise 2.6.4

$$|w|_{0;B_\rho} \leq N(K+1)^2\rho^2 \max_{B_{(K+1)\rho}} |f - f(0)| \leq N(K+1)^{2+\delta}\rho^{2+\delta}[f]_{\delta}.$$

By Theorem 2.5.2 (or Lemma 2.7.5) and Taylor's formula

$$|h - T_0^2 h|_{0;B_\rho} \le N\rho^3 [h]_{3;B_\rho} \le NK^{-3}|h|_{0;B_{(K+1)\rho}}$$

$$= NK^{-3}|h|_{0;\partial B_{(K+1)\rho}} = NK^{-3}|u - T_0^2 u|_{0;\partial B_{(K+1)\rho}}.$$

By Taylor's formula the last norm is less than $N(K+1)^{2+\delta}\rho^{2+\delta}[u]_{2+\delta}$. Therefore,

$$|h - T_0^2 h|_{0;B_\rho} \le NK^{-3}(K+1)^{2+\delta}\rho^{2+\delta}[u]_{2+\delta},$$

$$|u - T_0^2 u - T_0^2 h|_{0;B_\rho} \le |w|_{0;B_\rho} + |h - T_0^2 h|_{0;B_\rho}$$

$$\le N\rho^{2+\delta}\{(K+1)^{2+\delta}[f]_\delta + K^{-3}(K+1)^{2+\delta}[u]_{2+\delta}\}.$$

Since we can consider balls centered at any point in the same way, we get

$$[u]'_{2+\delta} \le N\{(K+1)^{2+\delta}[f]_\delta + K^{-3}(K+1)^{2+\delta}[u]_{2+\delta}\},$$

which by Theorem 3.3.1 implies that

$$[u]_{2+\delta} \le N(K+1)^{2+\delta}[f]_\delta + N_1 K^{-3}(K+1)^{2+\delta}[u]_{2+\delta},$$

and it remains only to take K so large that $N_1 K^{-3}(K+1)^{2+\delta} \le 1/2$. The theorem is proved.

THEOREM 3.4.2. *Let $k \ge 0$ be an integer. Then there exists a constant $N = N(d,\delta,k)$ such that for any $u \in C^{k+2+\delta}(\mathbb{R}^d)$ and any real λ we have*

$$[u]_{k+2+\delta} \le N\Big([\Delta u - \lambda^2 u]_{k+\delta} + |\lambda|^{k+\delta}|\Delta u - \lambda^2 u|_0\Big), \quad |\lambda|^2|u|_0 \le |\Delta u - \lambda^2 u|_0. \tag{3.4.2}$$

Proof. Estimates (3.4.2) for $\lambda = 0$ follow if we let $\lambda \to 0$. If $\lambda \ne 0$, the substitution $u(x) = v(\lambda x)$ reduces the situation to the case $\lambda = 1$ (we say a little bit more about it in the proof of Theorem 3.7.1 below). In this case our assertion means that

$$|u|_{k+2+\delta} \le N\Big([\Delta u - u]_{k+\delta} + |\Delta u - u|_0\Big). \tag{3.4.3}$$

By Theorem 3.4.1 for any α such that $|\alpha| = k$ we have

$$[D^\alpha u]_{2+\delta} \le N[\Delta(D^\alpha u)]_\delta \le N[\Delta u]_{k+\delta}.$$

It follows that

$$[u]_{k+2+\delta} \le N[\Delta u]_{k+\delta} \le N[\Delta u - u]_{k+\delta} + N[u]_{k+\delta}.$$

Furthermore, by Theorem 2.9.2 we have $|u|_0 \le |\Delta u - u|_0$, which by the above estimate and by the interpolation inequalities yields

$$|u|_{k+2+\delta} \le N[u]_{k+2+\delta} + N|u|_0 \le N[\Delta u - u]_{k+\delta} + N|u|_{k+\delta} + N|\Delta u - u|_0$$

$$\le N|\Delta u - u|_{k+\delta} + \frac{1}{2}|u|_{k+2+\delta} + N|u|_0 + N|\Delta u - u|_0$$

$$\le N\Big([\Delta u - u]_{k+\delta} + |\Delta u - u|_0\Big) + \frac{1}{2}|u|_{k+2+\delta}.$$

To get (3.4.3) it remains only to collect similar terms in the inequality between the extreme terms. The theorem is proved.

Theorems 3.4.2 and 2.8.2 allow us to solve Laplace's equation in the Hölder spaces.

THEOREM 3.4.3. *Let $f \in C^{k+\delta}(\mathbb{R}^d)$ and $\lambda \neq 0$. Then there exists a unique function $u \in C^{k+2+\delta}(\mathbb{R}^d)$ such that $\Delta u - \lambda^2 u = f$ in $\mathbb{R}^d$.*

This is a particular case of Theorem 3.7.2 below, whose proof would not become shorter if we considered only second–order operators.

EXERCISE* 3.4.4. Under the conditions of Theorem 3.4.2 prove that there exists a constant $N = N(d, \delta, k)$ such that for any $u \in C^{k+2+\delta}(\mathbb{R}^d)$ and any real λ and $r \in [0, k+2+\delta]$ we have

$$|\lambda|^{k+2+\delta-r}[u]_r \le N\Big([\Delta u - \lambda^2 u]_{k+\delta} + |\lambda|^{k+\delta}|\Delta u - \lambda^2 u|_0\Big).$$

EXERCISE 3.4.5 (cf. also Exercise 3.7.7 below). Prove that if $u \in C^2(\mathbb{R}^d)$, then $[u]_{1+\delta} \le N(|\Delta u|_0 + |u|_0)$, where N is independent of u. By using dilations conclude that $[u]_{1+\delta} \le N|\Delta u|_0^{(1+\delta)/2}|u|_0^{(1-\delta)/2}$. After referring to Theorem 2.9.2 prove that $[u]_{1+\delta} \le N|\Delta u - u|_0$ and, further, $|u|_{1+\delta} \le N|\Delta u - u|_0$.

EXERCISE 3.4.6. From Exercise 3.3.8 obtain that for $u \in C^{1+\delta}(\mathbb{R}^d)$ we have $[u]_{1+\delta} \le N\sup_{\varepsilon>0}\varepsilon[u^\varepsilon]_{2+\delta}$. Combine this with Theorem 3.4.1 and Exercise 3.2.10 to prove that if $p > d$ and we define δ by the equation $1 + \delta = 2 - d/p$ (so that $\delta \in (0,1)$) and $u \in C^{2+\delta}(\mathbb{R}^d)$, then $[u]_{1+\delta} \le N(d,p)||\Delta u||_{L_p}$.

EXERCISE 3.4.7. By using Exercise 3.3.8 for $n = 2$ and Exercise 3.2.10 prove that if $d/2 < p < d$ and we define δ by the equation $\delta = 2 - d/p$ (so that $\delta \in (0,1)$), then $[u]_\delta \le N(d,p)||\Delta u||_{L_p}$ for any $u \in C^{2+\delta}(\mathbb{R}^d)$.

3.5. An estimate for derivatives of L–harmonic functions

We return to the treatment of mth order elliptic equations $Lu = f$ $(m \ge 2)$. Here we will deal with solutions of the equation $Lu = 0$ in some domains. Such functions we call L-harmonic. The usefulness of the estimates we are talking about in this section is quite clear from Section 3.4.

For some purposes we need to introduce a real parameter λ into the equation. Thus let

$$L_\lambda = \sum_{|\alpha| \le m} a^\alpha \lambda^{m-|\alpha|} D^\alpha, \quad L_1 = L.$$

For $\lambda \neq 0$, by $G_\lambda(x)$ we denote the Green's function of L_λ. Notice that for $\lambda \neq 0$ and $v(x) = u(\lambda^{-1}x)$ we have $\lambda^m Lv(x) = [L_\lambda u](\lambda^{-1}x)$, and this along with the definition and the uniqueness of Green's functions shows at once that

$$G_\lambda(x) = \lambda^{d-m} G(\lambda x).$$

In turn this and Theorem 1.7.1 imply that for any multi–index α such that $d+|\alpha| \geq m+1$

$$|D^\alpha G_\lambda(x)| = \lambda^{d-m+|\alpha|}|(D^\alpha G)(\lambda x)| \leq \lambda^{d-m+|\alpha|} N \frac{1}{|\lambda x|^{d+|\alpha|-m}} = N \frac{1}{|x|^{d+|\alpha|-m}} \tag{3.5.1}$$

with a constant N independent of $\lambda \neq 0$ and $x \in \mathbb{R}^d \setminus \{0\}$. Also, by Corollary 1.5.1 we have $|D^\alpha G(x)| \leq N|x|^{-2r}$ whenever $2r \geq d+|\alpha|+1-m$. Therefore, for instance, for $|x| \geq 1$ and any α for all sufficiently large integers n, we have

$$|D^\alpha G_\lambda(x)| \leq \lambda^{d-m+|\alpha|} N \frac{1}{|\lambda x|^{n+d-m+|\alpha|}} \leq N \frac{1}{\lambda^n}, \tag{3.5.2}$$

where N depends, of course, on n but is independent of λ, x.

The following lemma is a more precise statement about solutions of $Lu = 0$ than Corollary 1.6.1.

LEMMA 3.5.1. *If $R > 0$ and a function $u \in C^m_{loc}(B_R) \cap C(B_R)$ and $Lu = 0$ in B_R, then u is infinitely differentiable in B_R and for any multi–index α there is a constant N independent of u and R such that*

$$|D^\alpha u(0)| \leq \frac{N}{R^{|\alpha|}} |u|_{0;B_R}. \tag{3.5.3}$$

Proof. First let $d + |\alpha| \geq m + 1$. Define $v(x) = u(Rx)$ and take a function $\zeta \in C_0^\infty(\mathbb{R}^d)$ such that

$$\zeta(x) = \begin{cases} 1 & \text{for } |x| \leq 1/2, \\ 0 & \text{for } |x| \geq 3/4. \end{cases}$$

Observe that $L_R v = 0$ in B_1. Therefore by applying formula (1.6.1) to L_R instead of L (and remembering that the coefficients of L_R contain R to appropriate powers) we get

$$R^{|\alpha|}|D^\alpha u(0)| = |D^\alpha v(0)|$$

$$= \Big| \sum_{\substack{|\gamma|+|\beta| \leq m \\ |\gamma| \geq 1}} c^{\gamma\beta} R^{m-|\gamma|-|\beta|} \int_{3/4>|y|>1/2} D^\alpha G_R(-y)[D^\gamma \zeta] D^\beta v(y)\, dy \Big|$$

$$= \Big| \sum_{\substack{|\gamma|+|\beta| \leq m \\ |\gamma| \geq 1}} b^{\gamma\beta} R^{m-|\gamma|-|\beta|} \int_{3/4>|y|>1/2} v(y)[D^{\beta+\alpha} G_R(-y)] D^\gamma \zeta(y)\, dy \Big|.$$

If $R \leq 2$, we use the inequality $d + |\beta + \alpha| \geq m + 1$ and (3.5.1). If $R \geq 2$, we use (3.5.2) with sufficiently large n. This yields

$$R^{|\alpha|}|D^\alpha u(0)| \leq N \max_{1/2 \leq |x| \leq 1} |v(x)| \leq N|u|_{0;B_R}.$$

For smaller values of $|\alpha|$ we use Lemma 3.1.4. For instance, if $d + |\alpha| = m$, then for a point $y \in B_{R/2}$ we have

$$|D^\alpha u(0)| \le |D^\alpha u(0) - D^\alpha u(y)| + |D^\alpha u(y)| \le NR \max_{j=1,\dots,d} \max_{z \in B_{R/2}} |D^{\alpha+e_j} u(z)|$$

$$+ \frac{N}{R^{|\alpha|}} |u|_{0;B_R} \le NR \max_{z\in B_{R/2}} \frac{1}{R^{|\alpha|+1}} |u|_{0;B_{R/2}(z)} + \frac{N}{R^{|\alpha|}} |u|_{0;B_R} \le \frac{N}{R^{|\alpha|}} |u|_{0;B_R}.$$

If needed, we can keep going to smaller values of $|\alpha|$ in the same way, and the lemma is proved.

REMARK 3.5.2. Actually our proof implies that

$$|D^\alpha u(0)| \le \frac{N}{R^{|\alpha|}} |u|_{0;B_R \setminus B_{R/2}}.$$

In the future we will also need the following elementary lemma showing, roughly speaking, that if the support of a bounded function f is small, then $G * f$ is small.

LEMMA 3.5.3. *If $d \ge 2$ and $\rho > 0$ and $|\alpha| = m-1$, then in B_ρ*

$$|D^\alpha G| * I_{B_\rho} \le N\rho, \quad |G| * \{(1+|\cdot|)I_{B_\rho}\} \le N\rho^{1+\delta}$$

where the constant N is independent of ρ.

Proof. For $x \in B_\rho$ we have

$$|D^\alpha G| * I_{B_\rho}(x) \le N \int_{|y|\le\rho} \frac{1}{|x-y|^{d-m+|\alpha|}}\, dy \le N \int_{|z|\le 2\rho} \frac{1}{|z|^{d-1}}\, dz = N\rho.$$

By using the rough estimate $(1+|y|)|G(y)| \le N|y|^{1+\delta-d}$ $(d \ge 1+\delta,\ m \ge 2)$ we conclude

$$|G| * \{(1+|\cdot|)I_{B_\rho}\}(x)| = \int_{|y|\le\rho} |G(x-y)|(1+|y|)\, dy$$

$$\le \int_{|y|\le\rho} |G(x-y)|\, dy + \rho \int_{|y|\le\rho} |G(x-y)|\, dy$$

$$\le N \int_{|z|\le 2\rho} \frac{1}{|z|^{d-1-\delta}}\, dz + N\rho \int_{|z|\le 2\rho} \frac{1}{|z|^{d-\delta}}\, dz = N\rho^{1+\delta}.$$

The lemma is proved.

EXERCISE 3.5.4. Denote by N_0 the smallest constant N which suits (3.5.3) for all $R > 0$ and unit α (and solutions u of $Lu = 0$). Prove that if $R > 0$ and a function $u \in C^m_{loc}(B_R) \cap C(B_R)$ and $Lu = 0$ in B_R, then for any multi–index α

$$|D^\alpha u(0)| \le \frac{N_0^k}{(R/k)^k} |u|_{0;B_R},$$

where $k = |\alpha|$. Conclude that u is real-analytic in B_R. Also prove again that $G(x)$ is real-analytic for $x \ne 0$.

EXERCISE 3.5.5. Let Ω be a connected domain. Prove that if $u \in C^m_{loc}(\Omega)$ and $Lu = 0$ in Ω and $u \equiv 0$ in a subdomain of Ω, then $u \equiv 0$ in Ω.

EXERCISE 3.5.6. In Lemma 3.5.1 replace L by L_λ and prove that (3.5.3) holds with N also independent of λ if $\lambda \neq 0$.

EXERCISE 3.5.7. Prove that the statement of Lemma 3.5.1 remains valid if we replace L by L_0 ($= L_\lambda$ for $\lambda = 0$).

EXERCISE 3.5.8. Show that if $u \in C^m_{loc}(\mathbb{R}^d)$, $|u(x)| \leq N(1 + |x|^n)$ with some constants N, n and $L_0 u = 0$, then u is a polynomial of degree less than or equal to n.

3.6. A priori estimates in the whole space for general elliptic operators

These estimates belong to Douglis and Nirenberg (1955), who actually treated systems of equations. We denote by L a mth order elliptic operator and take a number $\delta \in (0,1)$.

THEOREM 3.6.1. *There is a constant N such that for any $u \in C^{m+\delta}(\mathbb{R}^d)$ we have*

$$[u]_{m+\delta} \leq N\Big([Lu]_\delta + |u|_m\Big). \tag{3.6.1}$$

Proof. If $d = 1$, then

$$D^m u = (a^m)^{-1} Lu - (a^m)^{-1} \sum_{\alpha \leq m-1} a^\alpha D^\alpha u, \quad [u]_{m+\delta} \leq N\Big([Lu]_\delta + \sum_{\alpha \leq m-1} [u]_{\alpha+\delta}\Big)$$

$$\leq N\Big([Lu]_\delta + \sum_{\alpha \leq m-1} [u]_{\alpha+1}\Big) \leq N\Big([Lu]_\delta + |u|_m\Big).$$

In the remaining case $d \geq 2$ we use M. V. Safonov's idea of applying equivalent norms and representing functions as sums of "small" and " very smooth functions". To prove (3.6.1) take $|\alpha| = m - 1$ and denote

$$f = Lu, \quad g = L(T_0^1 D^\alpha u)$$

($T_y^k h$ is Taylor's expansion of h of order k centered at y) and use Remark 1.5.3. Upon observing that $0 \leq d - m + |\alpha| \leq d - 1$, so that $D^\alpha G$ is integrable, we obtain

$$w(x) := D^\alpha u(x) - T_0^1 D^\alpha u(x) = (D^\alpha G) * f(x) - G * g(x).$$

Also denote $f_0 = f(0)$ and notice that $(D^\alpha G) * f_0(x) = D^\alpha(G * f_0)(x) = D^\alpha \mathrm{const} = 0$. Therefore, $(D^\alpha G) * f(x) = (D^\alpha G) * (f - f_0)(x)$, and

$$w(x) = D^\alpha u(x) - T_0^1 D^\alpha u(x) = (D^\alpha G) * (f - f_0)(x) - G * g(x).$$

Next we represent the right-hand side as a sum of a "very smooth" function and a "small" function. Namely, for a constant $K \geq 1$ to be chosen later and any $\rho > 0$ we get

$$w(x) = [D^\alpha G * \{I_{B^c_{(K+1)\rho}}(f - f_0)\}(x) - G * \{I_{B^c_{(K+1)\rho}} g\}(x)]$$

$$+[D^\alpha G * \{I_{B_{(K+1)\rho}}(f - f_0)\}(x) - G * \{I_{B_{(K+1)\rho}} g\}(x)] =: h(x) + r(x).$$

Since $LD^\alpha G(x) = D^\alpha LG(x) = 0$ for $x \neq 0$ and we have bounds on derivatives of G, it holds that $Lh = 0$ in $B_{(K+1)\rho}$. By the mean value theorem and by Lemma 3.5.1, for $x \in B_\rho$

$$|h(x) - T_0^1 h(x)| \leq N\rho^2 [h]_{2;B_\rho} \leq N\rho^2 \frac{1}{(K\rho)^2} |h|_{0;B_{(K+1)\rho}} = N\frac{1}{K^2}|h|_{0;B_{(K+1)\rho}}$$

with N *independent* of K, u, ρ. Recall that $h = w - r$ and that from the mean value theorem we know that $|w|_{0;B_{(K+1)\rho}} \leq N(K+1)^{1+\delta}\rho^{1+\delta}[D^\alpha u]_{1+\delta}$. Then upon denoting $p = T_0^1 D^\alpha u + T_0^1 h$ we infer from the above that, for $x \in B_\rho$,

$$|D^\alpha u(x) - p(x)| = |w(x) - T_0^1 h(x)| \leq |h(x) - T_0^1 h(x)| + |r(x)|$$

$$\leq N\frac{1}{K^2}|h|_{0;B_{(K+1)\rho}} + |r(x)| \leq N\frac{1}{K^2}|w|_{0;B_{(K+1)\rho}} + N|r|_{0;B_{(K+1)\rho}},$$

$$|D^\alpha u(x) - p(x)| \leq N\frac{(K+1)^{1+\delta}}{K^2}\rho^{1+\delta}[u]_{m+\delta} + N|r|_{0;B_{(K+1)\rho}}. \tag{3.6.2}$$

To estimate the last term, we apply Lemma 3.5.3 (remember that $|\alpha| = m-1$ and $d \geq 2$). We also use $T_0^1 D^\alpha u(y) = D^\alpha u(0) + y^j D_j D^\alpha u(0)$, and

$$g(y) = a^0[D^\alpha u(0) + y^j D_j D^\alpha u(0)] + \sum_{j=1}^d a^{e_j} D_j D^\alpha u(0), \ |g(y)| \leq N|u|_m(1+|y|).$$

Then for $|x| \leq \rho$ (even for $|x| \leq (K+1)\rho$) we get

$$|D^\alpha G * \{I_{B_{(K+1)\rho}}(f - f_0)\}(x)|$$

$$\leq N(K+1)^\delta \rho^\delta [f]_\delta |D^\alpha G| * \{I_{B_{(K+1)\rho}}\}(x) \leq N(K+1)^{1+\delta}\rho^{1+\delta}[f]_\delta,$$

$$|G * \{I_{B_{(K+1)\rho}} g\}(x)| \leq N|u|_m |G| * \{(1+|\cdot|)I_{B_{(K+1)\rho}}\}(x) \leq N(K+1)^{1+\delta}\rho^{1+\delta}|u|_m.$$

Now from (3.6.2) we conclude

$$|D^\alpha u(x) - p(x)| \leq N\frac{(K+1)^{1+\delta}}{K^2}\rho^{1+\delta}[u]_{m+\delta} + N(K+1)^{1+\delta}\rho^{1+\delta}|u|_m$$

$$+N(K+1)^{1+\delta}\rho^{1+\delta}|f|_\delta, \quad E_1[D^\alpha u; B_\rho] \leq N\frac{(K+1)^{1+\delta}}{K^2}\rho^{1+\delta}[u]_{m+\delta}$$

$$+N(K+1)^{1+\delta}\rho^{1+\delta}|u|_m + N(K+1)^{1+\delta}\rho^{1+\delta}|f|_\delta.$$

We can certainly take any ball $B_\rho(x)$ instead of B_ρ, and the last estimate implies that

$$[D^\alpha u]'_{1+\delta} \leq N\frac{(K+1)^{1+\delta}}{K^2}[u]_{m+\delta} + N(K+1)^{1+\delta}|u|_m + N(K+1)^{1+\delta}|f|_\delta.$$

Finally use $[D^\alpha u]_{1+\delta} \leq N[D^\alpha u]'_{1+\delta}$, which yields

$$[u]_{m+\delta} \leq N_1 \frac{(K+1)^{1+\delta}}{K^2}[u]_{m+\delta} + N_2(K+1)^{1+\delta}|u|_m + N_3(K+1)^{1+\delta}|f|_\delta. \quad (3.6.3)$$

We emphasize that constants N_i here are independent of K, which allows us to take K so large that $N_1(K+1)^{1+\delta}K^{-2} \leq 1/2$, and then (3.6.3) leads us to (3.6.1). The theorem is proved.

3.7. Solvability of elliptic equations with constant coefficients

We take the operator L as in Section 3.6 and first prove an estimate similar to (3.6.1) but without the term $|u|_m$ on the right. For the operator $\Delta - \lambda^2$ this has been done in Theorem 3.4.2.

THEOREM 3.7.1. *Let $k \geq 0$ be an integer and $0 < \delta < 1$. Then there exists a constant N such that for any $u \in C^{k+m+\delta}(\mathbb{R}^d)$ and any real λ we have*

$$[u]_{k+m+\delta} + |\lambda|^{k+m+\delta}|u|_0 \leq N\Big([L_\lambda u]_{k+\delta} + |\lambda|^{k+\delta}|L_\lambda u|_0\Big). \quad (3.7.1)$$

Proof. Estimate (3.7.1) for $\lambda = 0$ follows if we let $\lambda \downarrow 0$. If $\lambda \neq 0$, then it is easy to check that for $v(x) = u(\lambda^{-1}x)$ we have

$$\lambda^m L_1 v(x) = [L_\lambda u](\lambda^{-1}x), \quad |\lambda|^m [L_1 v]_q = |\lambda|^{-q}[L_\lambda u]_q,$$

$$[v]_q = |\lambda|^{-q}[u]_q \quad \forall q \geq 0.$$

It follows that we need prove (3.7.1) only for $\lambda = 1$ and $L_\lambda = L$.

Next,

$$|u(x)| = \Big|\int_{\mathbb{R}^d} G(x-y)Lu(y)\,dy\Big| \leq |Lu|_0 \int_{\mathbb{R}^d} |G(y)|\,dy, \quad |u|_0 \leq N|Lu|_0.$$

This shows that it suffices only to prove that

$$[u]_{k+m+\delta} \leq N\Big([Lu]_{k+\delta} + |Lu|_0\Big). \quad (3.7.2)$$

By substituting $D^\alpha u$ with $|\alpha| = k$ instead of u in (3.6.1) we get

$$[D^\alpha u]_{m+\delta} \leq N[L(D^\alpha u)]_\delta + N|u|_{k+m} \leq N[Lu]_{k+\delta} + \frac{1}{2}[u]_{k+m+\delta}$$

$$+ N|u|_0 \leq N\Big([Lu]_{k+\delta} + |Lu|_0\Big) + \frac{1}{2}[u]_{k+m+\delta},$$

$$[u]_{k+m+\delta} \leq N\Big([Lu]_{k+\delta} + |Lu|_0\Big) + \frac{1}{2}[u]_{k+m+\delta}.$$

This implies (3.7.2), and the theorem is proved.

The a priori estimates allow us to get the following existence theorem.

THEOREM 3.7.2. *Let $\lambda \neq 0$, $k \geq 0$ be an integer and $0 < \delta < 1$. Then for any $f \in C^{k+\delta}(\mathbb{R}^d)$ there exists a unique solution $u \in C^{k+m+\delta}(\mathbb{R}^d)$ of the equation $L_\lambda u(x) = f(x)$, $x \in \mathbb{R}^d$.*

Proof. If in addition to our hypotheses $f \in C^{k+m+1}(\mathbb{R}^d)$, then by Theorem 1.5.2 (in the case of second–order operators one may prefer to use Theorem 2.8.2) there is a solution $u \in C^{k+m+1}(\mathbb{R}^d)$. By Theorem 3.7.1 and the interpolation inequalities we have

$$|u|_{k+m+\delta} \leq N|f|_{k+\delta} \tag{3.7.3}$$

with the constant N independent of f. Now take arbitrary $f \in C^{k+\delta}(\mathbb{R}^d)$ and construct the functions f^ε as in Exercise 3.2.3. Then for solutions u_ε of $L_\lambda u_\varepsilon = f^\varepsilon$ by (3.7.3) we get that the norms $|u_\varepsilon|_{k+m+\delta}$ are uniformly bounded. By Exercise 3.2.2 there is a subsequence $u_{\varepsilon(n)}$ with $\varepsilon(n) \to 0$ and a function $u \in C^{k+m+\delta}(\mathbb{R}^d)$ such that $D^\alpha u_{\varepsilon(n)} \to D^\alpha u$ if $|\alpha| \leq k+m$, and $f^{\varepsilon(n)} \to f$ at any point $x \in \mathbb{R}^d$. To show that the function u is the function we need, it remains only pass to the limit in the equation $L_\lambda u_{\varepsilon(n)} = f^{\varepsilon(n)}$. Since uniqueness follows from Theorem 3.7.1, our theorem is proved.

EXERCISE 3.7.3. For $d = 2, 0 < \delta < 1$ and $u \in C(R^2)$ prove that if $u_{xx}, u_{yy} \in C^\delta(\mathbb{R}^2)$, then $u_{xy} \in C^\delta(\mathbb{R}^2)$.

EXERCISE 3.7.4. Prove that if L is homogeneous elliptic (see the definition in Exercise 1.3.4) and its characteristic polynomial $\sum_{|\alpha|=m} i^{|\alpha|}a^\alpha\xi^\alpha$ is *real*, then for $0 < \delta < 1$ and integer $k \geq 0$ and $u \in C^{k+m+\delta}(\mathbb{R}^d)$ we have

$$[u]_{k+m+\delta} \leq N[Lu]_{k+\delta} \tag{3.7.4}$$

with the constant N independent of u. Conclude that, in particular,

$$[u]_{k+2+\delta} \leq N[\Delta u]_{k+\delta}. \tag{3.7.5}$$

EXERCISE 3.7.5. Forget all that we have obtained above relying on Green's functions or the maximum principle, but assume that the estimate (3.7.5) is known for any $\mathbb{R}^d$. By using only this estimate and the interpolation inequalities derive that for any $\delta \in (0, 1)$ and integer $k \geq 0$ and $u \in C^{k+2+\delta}(\mathbb{R}^d)$ and any real λ estimate (3.7.1) holds with N independent of u and λ and with $L_\lambda = \lambda^2 - \Delta$, $m = 2$.

EXERCISE* 3.7.6. Under the conditions of Theorem 3.7.1 prove that there exists a constant N such that for any $u \in C^{k+m+\delta}(\mathbb{R}^d)$ and any real λ and $r \in [0, k+m+\delta]$ we have

$$|\lambda|^{k+m+\delta-r}[u]_r \leq N\Big([L_\lambda u]_{k+\delta} + |\lambda|^{k+\delta}|L_\lambda u|_0\Big).$$

EXERCISE 3.7.7 (cf. Exercise 3.4.5). Prove that $[u]_{m-1+\delta} \leq N|Lu|_0$ if $u \in C^m(\mathbb{R}^d)$, where N is independent of u.

EXERCISE 3.7.8. In this and the next exercise we allow $m = 1$. Show that (3.7.4) holds for any $0 < \delta < 1$ and integer $k \geq 0$ and $u \in C_0^\infty(\mathbb{R}^d)$ with the constant N independent of u provided that L is homogeneous elliptic of order m (for instance the Cauchy–Riemann operator).

EXERCISE 3.7.9. Let L be a homogeneous elliptic operator. Show that there are no constants N, δ, k such that for any $u \in C_0^\infty(\mathbb{R}^d)$ we have $|u|_0 \leq N[Lu]_{k+\delta}$.

REMARK 3.7.10. We assumed that $m \geq 2$, but actually Theorem 3.7.1 is true for $m = 1$ as well. Indeed, in this case, as we know $d = 1$, and the estimate is obtained in a straightforward way.

3.8. Hints to exercises

3.2.5. Consider $(x - t)_+^\delta$ where t is a positive parameter.
3.2.6. Use the fact that ε in (3.2.1) is an arbitrary positive number.
3.2.7. Apply *the Hadamard formula*:

$$F(x,\xi) - F(x,\eta) = (\xi^i - \eta^i) \int_0^1 F_{\xi^i}(x, t\xi + (1-t)\eta)\, dt.$$

3.2.9. Consider $u_{x^i} - \delta_{h,i}u$ and notice that $\delta_{h,i}u(x) \leq u(x + he_j)/h$.
3.2.11. Take $s = 1$ in Exercise 3.2.6 and take ∇u instead of u.
3.3.4. To prove that $[u]''_\delta \leq N[u]_\delta$, observe that

$$(u^\varepsilon)_x(x) = \varepsilon^{-1} \int_{\mathbb{R}^d} \zeta_x(y)[u(x - \varepsilon y) - u(x)]\, dy.$$

To prove the converse inequality, consider $[u(x) - u^\varepsilon(x)] + [u^\varepsilon(x) - u^\varepsilon(y)] + [u^\varepsilon(y) - u(y)]$ and estimate the first and the last term through $[u]_\delta \varepsilon^\delta$, whereas to estimate in terms of $[u]''_\delta$ the middle term apply the mean value theorem. After this minimize with respect to ε.
3.3.6. To prove that $[u]''_r \leq N[u]_r$, write $r = k + \delta$ where k is an integer and $\delta \in (0,1]$. Next, take a multi-index α with $|\alpha| = n$ and write $\alpha = \beta + \gamma$ where $|\beta| = k$. Then observe that $|\gamma| = n - k > 0$ and

$$D^\alpha u^\varepsilon(x) = \varepsilon^{k-n} \int_{\mathbb{R}^d} D^\gamma \zeta(y)[D^\beta u(x - \varepsilon y) - D^\beta u(x)]\, dy.$$

To prove the converse inequality for $r = \delta$, apply Exercises 3.3.4 and 3.2.11 along with the observation that $[u^\varepsilon]_\delta \leq [u]_\delta$. After this apply the result to $D^\beta u$ instead of u.
3.3.8. If $n = 0$, the result follows from Exercise 3.2.3. If $n \geq 1$, to prove that $[u]''_\delta \leq N[u]_\delta$, proceed as in the hint to Exercise 3.3.4. To prove the converse, apply Exercises 3.3.6 with $r = \delta$ and also estimate $[u^\varepsilon]_n$ through $[u^\varepsilon]_{n+\delta}$ and $[u^\varepsilon]_\delta$ by Exercise 3.3.7, after which note that $[u^\varepsilon]_\delta \leq [u]_\delta$.
3.4.4. Apply (3.4.2) and the interpolation inequalities.
3.4.5. By imitating the proof of Theorem 3.4.1 prove that

$$\inf_{p \in \mathcal{P}_1} |u - p|_{0;B_\rho} \leq \rho^{1+\delta} N(K+1)^{1+\delta} K^{-2} [u]_{1+\delta} + (K+1)^2 \rho^2 |f|_0.$$

Use this inequality for $\rho \leq 1$, and for $\rho \geq 1$ use the fact that, obviously, $\inf_{p \in \mathcal{P}_1} |u - p|_{0;B_\rho} \leq \rho^{1+\delta} |u|_0$.
3.5.4. See Sec. 2.5 and the hint to Exercise 2.7.10.
3.5.7. One may assume that $R = 1$ and $d + |\alpha| \geq m + 1$. Also at first assume that u is infinitely differentiable in B_R and apply formula (1.6.1) for L_λ instead of L with small $\lambda \neq 0$. Then obtain an estimate on $D^\alpha u(0)$ in which let $\lambda \to 0$. Finally consider $u * \phi$ instead of u.
3.7.3. At a certain stage you may need to use uniqueness of solutions to differential equations. See also Theorem 5.1.3 below.

3.7.4. Consider either $Lu+u$ or $Lu-u$.
3.7.5. Consider $u(x)\exp(i\lambda x^0)$ where x^0 is a new independent variable.
3.7.6. Apply (3.7.1) and the interpolation inequalities.
3.7.7. Take the function ζ from Exercise 3.2.3 and observe that for $\varepsilon,\delta_1\in(0,1)$ we have

$$[u^{\varepsilon}]_{m+\delta_1}\le N[(Lu)^{\varepsilon}]_{\delta_1}\le N|Lu|_0\varepsilon^{-\delta_1}.$$

This by the interpolation inequalities yields $[u^{\varepsilon}]_m\le N|Lu|_0\varepsilon^{-\delta_1 m/(m+\delta_1)}$. Then use Exercise 3.3.4.
3.7.8. Use Exercise 3.7.4 for $L\bar{L}^*$ instead of L and then Exercise 1.3.4 for $\bar{L}^*$ instead of L after noticing that $[\bar{L}^*u]_{k+m+\delta}\le N[u]_{k+2m+\delta}$.
3.7.9. Use dilations.

CHAPTER 4

Elliptic Equations with Variable Coefficients in $\mathbb{R}^d$

In this chapter we start a systematic study of elliptic equations with variable coefficients. So let $a^\alpha(x)$ be some (complex) functions given for $|\alpha| \leq m$, $x \in \mathbb{R}^d$. Assume that for a constant $\kappa > 0$ and all $\xi, x \in \mathbb{R}^d$ we have

$$| \sum_{|\alpha| \leq m} a^\alpha(x) i^{|\alpha|} \xi^\alpha | \geq \kappa(1 + |\xi|^m),$$

so that the operator $L = L(x) = \sum_{|\alpha| \leq m} a^\alpha(x) D^\alpha$ is *uniformly* elliptic. For real λ we also define

$$L_\lambda = \sum_{|\alpha| \leq m} a^\alpha(x) \lambda^{m-|\alpha|} D^\alpha.$$

Finally we fix a $\delta \in (0, 1)$.

4.1. Schauder's a priori estimates

Our first step is to obtain a priori estimates in the Hölder spaces. For elliptic operators with variable coefficients they are called Schauder's estimates. If one is only interested in second-order elliptic operators, the corresponding estimates can be obtained in a way which may look somewhat easier than in the general setting. We refer the interested reader to the proofs of Lemma 8.9.1 and Theorem 8.9.2 and to Exercise 8.9.3. These proofs can be repeated literally (almost) for elliptic equations. The main reason for the distinction of proofs for second-order and general operators is due to the fact that in the case of second-order operators we are not afraid of getting $|u|_0$ on the right in our estimates, since from the maximum principle we know how to estimate it through $|Lu|_0$. Finally, note that we need the presence of the parameter λ even for second-order equations in order to investigate these equations in domains, whereas admission of complex coefficients allows us in Secs. 8.2 and 8.3 to reduce solving parabolic equations to solving elliptic ones.

We need the following localization property of the spaces $C^{n+\delta}$.

LEMMA 4.1.1. *Fix a cut-off function $\zeta \in C_0^\infty(\mathbb{R}^d)$ such that $\zeta(x) = 1$ for $|x| \leq 1$ and $\zeta(x) = 0$ for $|x| \geq 2$ and $0 \leq \zeta \leq 1$, and fix an integer n. Then there is a constant $N = N(d, n, \theta)$ such that for any $\theta \in [0, 1]$, $R \geq 1$ and $u \in C^{n+\theta}(\mathbb{R}^d)$ we have*

$$|u|_{n+\theta} \leq N \sup_{y \in \mathbb{R}^d} \Big([u\zeta_R^y]_{n+\theta} + |u\zeta_R^y|_0 \Big),$$

where $\zeta_R^y(x) = \zeta(R^{-1}(x - y))$.

Proof. First we claim that for any integer r and $\tau \in [0,1]$

$$[u]_{r+\tau} \le 2 \sup_{y\in\mathbb{R}^d} \Big([u]_{r+\tau;B_1(y)} + [u]_{r;B_1(y)}\Big). \tag{4.1.1}$$

This is obvious if $\tau = 0$ or 1. If $\tau \in (0,1)$, then we let $|\alpha| = r$ and consider the quotient

$$\frac{|D^\alpha u(x_1) - D^\alpha u(x_2)|}{|x_1 - x_2|^\tau}.$$

If both x_1, x_2 are in some ball of radius 1, then clearly the quotient is less than the right–hand side of (4.1.1). If $|x_1 - x_2| \ge 2$, then the quotient is less than $2\sup_x |D^\alpha u(x)|$, which is again less than the right–hand side of (4.1.1). This proves (4.1.1).

Now fix y and observe that for $x \in B_1(y)$

$$u(x) = cR^{-d} \int_{|z-y|\le 3R} u(x)\zeta(R^{-1}(x-z))\,dz, \quad c^{-1} := \int_{\mathbb{R}^d} \zeta(z)\,dz.$$

Therefore,

$$[u]_{r+\tau;B_1(y)} \le cR^{-d} \int_{|z-y|\le 3R} [u\zeta^z_R]_{r+\tau}\,dz \le N \sup_{z\in\mathbb{R}^d} [u\zeta^z_R]_{r+\tau}.$$

Hence and by (4.1.1) and the interpolation inequalities for $r+\tau \le n+\theta$

$$[u]_{r+\tau} \le N \sup_{y\in\mathbb{R}^d} \Big([u\zeta^y_R]_{r+\tau} + |u\zeta^y_R|_r\Big) \le N \sup_{y\in\mathbb{R}^d} \Big([u\zeta^y_R]_{n+\theta} + |u\zeta^y_R|_0\Big).$$

Taking here $\tau = 0, r = 0, ..., n$ and $r+\tau = n+\theta$, we get our assertion. The lemma is proved.

THEOREM 4.1.2. *Assume that for any α we have $|a^\alpha|_\delta \le K$ where K is a constant. Then there exist constants $N, \lambda_0 \ge 0$ depending only on κ, m, δ, K, d, such that for any $u \in C^{m+\delta}(\mathbb{R}^d)$ and any real λ with $|\lambda| \ge \lambda_0$ we have*

$$[u]_{m+\delta} + |\lambda|^{m+\delta}|u|_0 \le N\Big([L_\lambda u]_\delta + |\lambda|^\delta |L_\lambda u|_0\Big). \tag{4.1.2}$$

Proof. If $L_\lambda u = f$, then for the function $v(x) := u(x\lambda^{-1})$ we have

$$\sum_{|\alpha|\le m} a^\alpha(x\lambda^{-1}) D^\alpha v(x) = \lambda^{-m} f(x\lambda^{-1}).$$

Moreover, $[a^\alpha(\lambda^{-1}\cdot)]_\delta = |\lambda|^{-\delta}[a^\alpha]_\delta$. It follows easily that we need to prove (4.1.2) only for $\lambda = 1$ under the assumption that $[a^\alpha]_0 \le K$ and $[a^\alpha]_\delta$ are less than a certain small positive constant depending only on κ, m, K, δ, d. To fix the setting, we assume that $[a^\alpha]_\delta \le \gamma$, where γ is a small constant to be specified later.

Take a constant $R \ge 1$, the function ζ from Lemma 4.1.1, and take a point $y \in \mathbb{R}^d$. Then by Theorem 3.7.1 we have

$$|u\zeta^y_R|_{m+\delta} \le N\Big([L(y)(u\zeta^y_R)]_\delta + |L(y)(u\zeta^y_R)|_0\Big). \tag{4.1.3}$$

Usually one says that (4.1.3) is obtained by *freezing* the coefficients of L.

After evaluating $L(y)(u(x)\zeta_R^y(x))$ and applying (3.1.6) we see that

$$[L(y)(u\zeta_R^y)]_\delta \le [\zeta_R^y L(y)u]_\delta + NR^{-1}|u|_{m-1+\delta}$$

$$\le [\zeta_R^y Lu]_\delta + [\zeta_R^y \{L(y) - L\}u]_\delta + NR^{-1}|u|_{m+\delta}.$$

Moreover,

$$[\zeta_R^y Lu]_\delta \le [Lu]_\delta + NR^{-\delta}|u|_m,$$

$$[\zeta_R^y\{L(y) - L\}u]_\delta \le N \max_{|\alpha|\le m} |\zeta_R^y\{a^\alpha(y) - a^\alpha\}|_0|u|_{m+\delta}$$

$$+N|u|_m\{[\zeta_R^y]_\delta + \max_{|\alpha|\le m}[a^\alpha]_\delta\} \le N(\gamma R^\delta + R^{-\delta} + \gamma)|u|_{m+\delta}.$$

Similarly

$$|L(y)(u\zeta_R^y)|_0 \le |\zeta_R^y Lu|_0 + |\zeta_R^y\{L(y) - L\}u|_0 + NR^{-1}|u|_m$$

$$\le |Lu|_0 + N\gamma R^\delta|u|_m + NR^{-1}|u|_m.$$

Coming back to (4.1.3), we get

$$|u\zeta_R^y|_{m+\delta} \le N|Lu|_\delta + N(R^{-\delta} + \gamma R^\delta + \gamma)|u|_{m+\delta},$$

which by Lemma 4.1.1 implies that

$$|u|_{m+\delta} \le N|Lu|_\delta + N_1(R^{-\delta} + \gamma R^\delta + \gamma)|u|_{m+\delta}.$$

It remains only to observe that one can choose first R and then γ so that

$$N_1(R^{-\delta} + \gamma R^\delta + \gamma) \le 1/2.$$

The theorem is proved.

EXERCISE 4.1.3. Assume that (4.1.2) is satisfied for $\lambda = \lambda_1$ whenever the assumptions of Theorem 4.1.2 are satisfied. Prove then that (4.1.2) also holds for $\lambda = t\lambda_1$ with any real t such that $|t| \ge 1$.

EXERCISE 4.1.4. Take the function $F(x,\xi)$ as in Exercise 3.2.7 and assume that $n \le 2, k < 1$. Prove that there is a constant N and a finite function $N(s)$ such that for any $u \in C^{2+\delta}(\mathbb{R}^d)$ we have

$$|u|_{2+\delta} \le N[\Delta u + F(\cdot, u_x)]_\delta + N(|u|_0).$$

4.2. Better regularity of Lu implies better regularity of u

In this section we show that if Lu is in a better class than only in C^δ, then u also belongs to a better class than $C^{m+\delta}(\mathbb{R}^d)$. The arguments in this section are independent of the order of operators, so that considering only second–order operators would not simplify the presentation. However, the reader only interested in $C^{2+\delta}$–theory of second–order elliptic equations may skip this section.

THEOREM 4.2.1. *Let the assumptions of Theorem 4.1.2 be satisfied, and let $k \geq 0$ be an integer, $K_1 \geq 1$ be a constant. Assume that for any α we have $|a^\alpha|_{k+\delta} \leq K_1$. Then for any λ, the inclusions $u \in C^{m+\delta}(\mathbb{R}^d)$ and $L_\lambda u \in C^{k+\delta}(\mathbb{R}^d)$ imply that $u \in C^{k+m+\delta}(\mathbb{R}^d)$. Moreover, if we take λ_0 from Theorem 4.1.2 and take a real λ so that $|\lambda| \geq \lambda_0$, then there exists a constant $N > 0$ depending only on $\kappa, k, m, \delta, K_1, d$, such that for any $u \in C^{k+m+\delta}(\mathbb{R}^d)$ we have*

$$[u]_{k+m+\delta} + |\lambda|^{k+m+\delta}|u|_0 \leq N\Big([L_\lambda u]_{k+\delta} + |\lambda|^{k+\delta}|L_\lambda u|_0\Big). \tag{4.2.1}$$

Proof. For $k = 0$ this is just Theorem 4.1.2. Therefore, we can use induction on k. Assume that our theorem is true for $k = k_0$ and its assumptions are satisfied for $k = k_0 + 1$, and take a function $u \in C^{m+\delta}(\mathbb{R}^d)$ such that for a λ we have $L_\lambda u \in C^{k_0+1+\delta}(\mathbb{R}^d)$. Observe at once that by the assumption actually $u \in C^{k_0+m+\delta}(\mathbb{R}^d)$. In turn, since the difference $L_\mu u - L_\lambda u$ is a sum of products of functions of class $C^{k_0+1+\delta}(\mathbb{R}^d)$ times derivatives of u of order less than or equal to $m-1$, the latter being of class $C^{k_0+1+\delta}(\mathbb{R}^d)$, it follows that the difference belongs to $C^{k_0+1+\delta}(\mathbb{R}^d)$. In particular, $L_\lambda u \in C^{k_0+1+\delta}(\mathbb{R}^d)$ for those λ for which estimate (4.2.1) is true for $k = k_0$.

Fix such a λ and take $h > 0$, $j = 1, ..., d$ and observe that the formula $\delta_{h,j}(vu)(x) = v(x)\delta_{h,j}u(x) + u(x+he_j)\delta_{h,j}v(x)$ implies that

$$L_\lambda(\delta_{h,j}u)(x) = \delta_{h,j}L_\lambda u(x) - \sum_{|\alpha|\leq m} \lambda^{m-|\alpha|}[D^\alpha u(x+he_j)]\delta_{h,j}a^\alpha(x). \tag{4.2.2}$$

Furthermore, for any differentiable function v

$$\delta_{h,j}v(x) = \int_0^1 \frac{\partial v}{\partial x^j}(x+the_j)\,dt,$$

which implies that $|\delta_{h,j}v|_{k+\delta} \leq |v|_{k+1+\delta}$. Hence, $|\delta_{h,j}v|_{k+\delta}$ is bounded as a function of h whenever $v \in C^{k+1+\delta}(\mathbb{R}^d)$. It follows that the $C^{k_0+\delta}(\mathbb{R}^d)$-norm of the right–hand side of (4.2.2) is bounded as a function of h. By the assumption from (4.2.1) and from the interpolation inequalities we get that also $|\delta_{h,j}u|_{k_0+m+\delta}$ is bounded as a function of h. Of course, this shows that $D_j u \in C^{k_0+m+\delta}(\mathbb{R}^d)$, and since j is arbitrary, $u \in C^{k_0+1+m+\delta}(\mathbb{R}^d)$.

It remains to carry estimate (4.2.1) from the case $k = k_0$ over the case $k = k_0+1$. Actually here we do not even need the assumption that the estimate is true for $k = k_0$. We will just prove it for any $u \in C^{k+m+\delta}(\mathbb{R}^d)$.

Bearing in mind the possibility of dilation of the space, we may and therefore will consider only $\lambda = \lambda_0$. To simplify notation we further assume that $\lambda_0 = 1$.

Take any multi–index α with $|\alpha| = k$ and apply Theorem 4.1.2. Then

$$|D^\alpha u|_{m+\delta} \leq N\Big([LD^\alpha u]_\delta + |LD^\alpha u|_0\Big) \leq N[LD^\alpha u]_\delta + N|u|_{k+m}.$$

Here

$$LD^\alpha u = D^\alpha Lu + \sum_{\substack{|\eta|\le m, |\beta|\ge 1 \\ |\beta+\sigma|=k}} c^{\beta\sigma}[D^\beta a^\eta] D^{\eta+\sigma} u,$$

where $c^{\beta\sigma}$ are some constants. Since $|\eta+\sigma| \le m + (k-1)$ and $|\beta| \le k$, it follows that

$$[LD^\alpha u]_\delta \le [D^\alpha Lu]_\delta + N|u|_{k+m-1+\delta}, \quad |D^\alpha u|_{m+\delta} \le N[Lu]_{k+\delta} + N|u|_{k+m},$$

$$[u]_{k+m+\delta} \le N[Lu]_{k+\delta} + N|u|_{k+m}. \tag{4.2.3}$$

Finally we use the fact that by the interpolation inequalities $|u|_{k+m} \le \varepsilon[u]_{k+m+\delta} + N(\varepsilon)|u|_0$, so that we can replace $N|u|_{k+m}$ on the right in (4.2.3) by $N|u|_0$. Then it remains only to observe that as we know from Theorem 4.1.2, $|u|_0 \le N\Big([Lu]_\delta + |Lu|_0\Big)$, and $[Lu]_\delta \le N\Big([Lu]_{k+\delta} + |Lu|_0\Big)$. The theorem is proved.

REMARK 4.2.2. In Theorems 7.1.1 and 7.1.2 we give a "local" version of Theorem 4.2.1.

EXERCISE* 4.2.3. Let $|a^\alpha|_{k+\delta} \le K_1$. Prove that there exists a constant $N = N(d, m, \kappa, K_1, k, \delta)$ such that for any $u \in C^{k+m+\delta}$ we have

$$|u|_{k+m+\delta} \le N\Big([L_0 u]_{k+\delta} + |u|_0\Big), \quad |u|_{k+m+\delta} \le N\Big([Lu]_{k+\delta} + |u|_0\Big).$$

EXERCISE 4.2.4 (better regularity of solutions of nonlinear equations). For a smooth real-valued function u, the set of its derivatives $D^\alpha u$, $|\alpha| \le m$, can be considered as a point in a Euclidean space of appropriate dimension. We denote this space by $\mathbb{R}^{d_1}$, and by common abuse of notation its generic point will be represented as $\xi^{(\alpha)}$ instead of $(\xi^{(\alpha)}, |\alpha| \le m)$. Let $F(x, \xi^{(\alpha)})$ be a real-valued $C^{1+\delta}(\mathbb{R}^d \times \mathbb{R}^{d_1})$-function, and let $u \in C^{m+\delta}(\mathbb{R}^d)$ be a solution of the equation $F(x, D^\alpha u(x)) = 0$ on $\mathbb{R}^d$. Assume that the equation is *elliptic on* u, which means that if we define $a^\beta(x) = F_{\xi^{(\beta)}}(x, D^\alpha u(x))$ and $L = \sum_{|\alpha|\le m} a^\alpha D^\alpha$, then L is an elliptic operator. Prove that in this situation $u \in C^{1+m+\delta}(\mathbb{R}^d)$. What happens if F is a $C^{k+\delta}(\mathbb{R}^d \times \mathbb{R}^{d_1})$-function?

4.3. Solvability of second–order elliptic equations with variable coefficients. The method of continuity

In this and the following sections we show the strength of a priori estimates by proving solvability of elliptic equations in the whole space. Here we show how the method of continuity works for second–order elliptic operators. Also we consider only operators with *real* coefficients. The operators under consideration fall into the scheme of general elliptic operators investigated above. But in this simplest situation we prefer to introduce new notation and state our assumptions somewhat differently. Specifically, our operators look like

$$L = a^{ij}(x)\frac{\partial^2}{\partial x^i \partial x^j} + b^i(x)\frac{\partial}{\partial x^i} + c(x),$$

where $a^{ij}(x), b^i(x), c(x)$ are real–valued functions. Assume that for a constant $\kappa > 0$ and all $x, \xi \in \mathbb{R}^d$

$$a^{ij}(x)\xi^i\xi^j \geq \kappa|\xi|^2, \quad c(x) \leq 0.$$

The constant κ is called *the ellipticity constant* of L. Without loss of generality we assume that the matrix a is symmetric. Finally notice that in this section we consider only real–valued solutions.

The reader who did not follow Sec. 4.2 must take $k = 0$ below and use the results from Sec. 4.1 instead of the results from Sec. 4.2.

THEOREM 4.3.1. *Take an integer $k \geq 0$ and assume that $a^{ij}, b^i, c \in C^{k+\delta}(\mathbb{R}^d)$ for any i, j. Then $u \in C^{k+2+\delta}(\mathbb{R}^d)$ whenever $u \in C^{2+\delta}(\mathbb{R}^d)$ and $Lu - zu \in C^{k+\delta}(\mathbb{R}^d)$ for a $z \in \mathbb{R}$. Moreover, for any $z > 0$ and any $u \in C^{k+2+\delta}(\mathbb{R}^d)$ we have*

$$[u]_{k+2+\delta} \leq N\Big([f]_{k+\delta} + (z^{-1} + z^{(k+\delta)/2})|f|_0\Big), \quad z|u|_0 \leq |f|_0, \tag{4.3.1}$$

where $f := Lu - zu$ and the constant N depends only on d, κ, δ and norms of a, b, c in $C^{k+\delta}(\mathbb{R}^d)$.

Proof. The second estimate in (4.3.1) follows from Theorem 2.9.2. Next, let $L_0 u := a^{ij}D_iD_ju$. One sees easily that the operator $L_0 - 1$ is an elliptic operator to which Theorem 4.2.1 can be applied. Therefore, for instance, $u \in C^{k+2+\delta}(\mathbb{R}^d)$ whenever $u \in C^{2+\delta}(\mathbb{R}^d)$ and $L_0 u \in C^{k+\delta}(\mathbb{R}^d)$. We can replace here the operator L_0 with $L - z$ in the same way as in the proof of Theorem 4.2.1 after noticing that their difference is a lower–order operator with regular coefficients. This proves the first statement.

Also by Theorem 4.2.1 there is $z_0 \geq 0$, depending only on d, κ, δ and norms of a, b, c in $C^\delta(\mathbb{R}^d)$, such that

$$[u]_{k+2+\delta} \leq N\Big([L_0u - z_0u]_{k+\delta} + |L_0u - z_0u|_0\Big) \tag{4.3.2}$$

with N depending only on d, κ, δ and norms of a, b, c in $C^{k+\delta}(\mathbb{R}^d)$. Furthermore, by the interpolation inequalities

$$[(L_0 - L)u - (z_0 - z)u]_{k+\delta} + |(L_0 - L)u - (z_0 - z)u|_0 \leq$$

$$N(|u|_{k+1+\delta} + (z+1)[u]_{k+\delta} + (z+1)|u|_0) \leq \gamma[u]_{k+2+\delta} + N(\gamma)(z+1)^{(k+\delta)/2+1}|u|_0),$$

where $\gamma > 0$ is any number and the constant $N(\gamma)$ also depends on other quantities under control. This and (4.3.2) for an appropriate choice of γ yield

$$[u]_{k+2+\delta} \leq (1/2)[u]_{k+2+\delta} + N\Big([f]_{k+\delta} + |f|_0 + (z+1)^{(k+\delta)/2+1}|u|_0\Big),$$

and we get the first inequality in (4.3.1) after using the second one along with the observation that $1 + (z+1)^{(k+\delta)/2+1}z^{-1} \leq N(z^{-1} + z^{(k+\delta)/2})$. The theorem is proved.

THEOREM 4.3.2. *Under the hypotheses of Theorem 4.3.1, for any $z > 0$ and $f \in C^{k+\delta}(\mathbb{R}^d)$ there exists a unique solution $u \in C^{k+2+\delta}(\mathbb{R}^d)$ of the equation $Lu(x) - zu = f(x)$, $x \in \mathbb{R}^d$.*

Proof. The uniqueness follows at once from Theorem 4.3.1, which also implies that we need consider only the case $k = 0$. We apply the *method of continuity*. For $t \in [0, 1]$ define

$$L^t = tL + (1 - t)\Delta = (ta^{ij} + (1 - t)\delta^{ij})\frac{\partial^2}{\partial x^i \partial x^j} + tb^i\frac{\partial}{\partial x^i} + tc,$$

and let T be the set of all points $t \in [0, 1]$ for which the statement of the theorem is true (with L^t in place of L). By the above $0 \in T$, so that T is not empty. Obviously estimate (4.3.1) holds with the same constant N for all solutions of the equations $L^t u - zu = f$. This immediately implies that T is closed.

Therefore, to finish the proof it remains only to prove that T is open in the topology of $[0, 1]$. Take any point $t_0 \in T \cap [0, 1]$ and define the linear operator $\mathcal{R} : C^\delta(\mathbb{R}^d) \to C^{2+\delta}(\mathbb{R}^d)$ such that it takes any $f \in C^\delta(\mathbb{R}^d)$ into the solution of $L^{t_0}u - zu = f$. By the assumption $\mathcal{R}$ is well defined and by Theorem 4.3.1 is bounded. Now in order to show that for $t \in [0, 1]$ close to t_0 the equation $L^t u - zu = f$ is solvable we write this equation as

$$L^{t_0}u - zu = f + (L^{t_0} - L^t)u, \quad u = \mathcal{R}f + \mathcal{R}(L^{t_0} - L^t)u,$$

and we show that the operator $\mathcal{R}(L^{t_0} - L^t)$ is a contraction in $C^{2+\delta}(\mathbb{R}^d)$ for *all* t close to t_0.

By the above for certain constants N independent of t, u

$$|\mathcal{R}(L^{t_0} - L^t)u|_{2+\delta} \leq N|(L^{t_0} - L^t)u|_\delta$$

$$= N|t_0 - t| \cdot |(L - \Delta)u|_\delta \leq N_1|t_0 - t| \cdot |u|_{2+\delta}.$$

For t such that $N_1|t_0 - t| \leq 1/2$ the operator $\mathcal{R}(L^{t_0} - L^t)$ is indeed a contraction, and the theorem is proved.

REMARK 4.3.3. Actually the argument based on closeness and openness of T is needed only in situations more complicated than the one in Theorem 4.3.2. In the above case the constant N_1 is independent of t_0, and one can reach the point $t = 1$ starting with $t = 0$ in a finite number of steps defining $t_0 = 0$ and t_{i+1} so that $N_1|t_{i+1} - t_i| \leq 1/2$.

COROLLARY 4.3.4. *If a, b, c, f are infinitely differentiable and their every derivative of any order is bounded, the same holds for the solution u of the equation $Lu - zu = f$.*

This corollary can be localized; see Theorems 7.1.1 and 7.1.2.

REMARK 4.3.5. Take an integer $k \geq 0$ and a constant $K > 0$. Theorems 4.3.1 and 4.3.2 mean that if $|a, b, c|_{k+\delta} \leq K$ and $z > 0$, then the operator

$$z - L : C^{k+2+\delta}(\mathbb{R}^d) \to C^{k+\delta}(\mathbb{R}^d)$$

is onto and one-to-one. Moreover, if we denote by $\mathcal{R}_z$ its inverse operator, then for any $z_1 > 0$ there is a constant N depending only on $\kappa, \eta, \delta, K, d, z_1$ such that for $f \in C^{k+\delta}(\mathbb{R}^d)$ and $z \geq z_1$

$$[\mathcal{R}_z f]_{k+2+\delta} \leq N z^{(k+\delta)/2}|f|_{k+\delta}, \quad z|\mathcal{R}_z f|_0 \leq |f|_0.$$

From the multiplicative inequality $[u]_s \le N[u]_r^{s/r}|u|_0^{1-s/r}, 0 \le s \le r$, we also get that for $r \le k+2+\delta$

$$[\mathcal{R}_z f]_r \le N z^{r/2-1}|f|_{k+\delta}, \quad |\mathcal{R}_z f|_r \le N z^{r/2-1}|f|_{k+\delta}.$$

In particular, we have

$$|\mathcal{R}_z f|_2 \le N|f|_\delta, \quad |\mathcal{R}_z f|_1 \le |z|^{-1/2} N|f|_\delta,$$

$$|\mathcal{R}_z f|_{1+\delta} \le |z|^{(\delta-1)/2} N|f|_\delta, \quad |\mathcal{R}_z f|_\delta \le |z|^{-1+\delta/2} N|f|_\delta.$$

Also Corollary 2.9.3 implies that the operators $\mathcal{R}_z$ are independent of k and δ.

EXERCISE 4.3.6. Let $F(\xi)$ be a continuously differentiable real–valued function on $\mathbb{R}^d$ such that $F(0) = 0$. Prove that for any $u \in C^2(\mathbb{R}^d)$ we have $|u|_0 \le |f|_0$, where $f := \Delta u + F(\operatorname{grad} u) - u$. Also prove that for any $g \in C(\mathbb{R}^d)$ any two solutions of class $C^2(\mathbb{R}^d)$ of the equation $\Delta u + F(\operatorname{grad} u) - u = g$ coincide.

EXERCISE 4.3.7. Let $F(\xi)$ be a twice continuously differentiable real-valued function on $\mathbb{R}^d$. Prove that for any $u \in C^{3+\delta}(\mathbb{R}^d)$ we have $|u|_{3+\delta} \le \Psi(|f|_{1+\delta})$, where $f := \Delta u + F(\operatorname{grad} u) - u$ and the function Ψ is a finite function independent of u.

EXERCISE 4.3.8. Take F from Exercise 4.3.6 and assume that

$$|F| \le |\xi|^{k+1}, \quad |F_\xi| \le |\xi|^k \tag{4.3.3}$$

with $k \in [0,1)$. By using Exercises 4.3.6 and 4.1.4 prove that for any number $M > 0$ there exists a constant N such that for any $f \in C^\delta(\mathbb{R}^d)$ with $|f|_\delta \le M$ and any function $\Phi(\xi)$, which coincides with F for $|\xi| \le N$ and satisfies (4.3.3), the equations

$$\Delta u + F(\operatorname{grad} u) - u = f, \quad \Delta u + \Phi(\operatorname{grad} u) - u = f$$

about $u \in C^{2+\delta}(\mathbb{R}^d)$ are equivalent. Also prove that if $k = 0$ and $u \in C^{2+\delta}(\mathbb{R}^d)$, then $|u|_{2+\delta} \le N(|\Delta u + F(\operatorname{grad} u) - u|_\delta)$ where $N = N(s)$ is a finite function independent of u.

EXERCISE 4.3.9. Take F from Exercise 4.3.6 and assume that F is Lipschitz continuous (in particular, F satisfies (4.3.3) with $N|\xi|, N$ instead of $|\xi|^{k+1}, |\xi|^k$). Prove that for any $f \in C^\delta(\mathbb{R}^d)$ there exists a unique solution $u \in C^{2+\delta}(\mathbb{R}^d)$ of the equation $\Delta u + F(\operatorname{grad} u) - u = f$.

EXERCISE 4.3.10. By using the previous exercises prove that for any $f \in C^\delta(\mathbb{R}^d)$ there exists a unique solution $u \in C^{2+\delta}(\mathbb{R}^d)$ of the equation $\Delta u + |\operatorname{grad} u|^{k+1} - u = f$, where $k \in [0,1)$. (Warning: The case $k = 0$ may present some difficulties.)

EXERCISE 4.3.11. Take the function F from Exercise 4.3.7 and prove that for any $f \in C^{1+\delta}(\mathbb{R}^d)$ there exists a unique $u \in C^{3+\delta}(\mathbb{R}^d)$ solving the equation $\Delta u + F(\operatorname{grad} u) - u = f$ in $\mathbb{R}^d$.

4.4. The case of second–order equations $Lu - zu = f$ with z complex

Our investigation of elliptic operators with complex–valued coefficients allows us to obtain an important and nontrivial information for second–order equations even if only the zeroth order term is complex–valued. Of course, we consider complex–valued solutions. Later the results of this section will be used only in investigation of the Cauchy problem for parabolic equations by means of semigroups in Secs. 8.2 and 8.3.

We take a $\delta \in (0,1)$ and an operator L from the previous section. We also fix a constant $\eta \in [0,\pi)$ and define

$$E = E_\eta = \{z = a + ib \in C \setminus \{0\} : |\arg z| \le \eta\}.$$

Let us first rewrite Theorem 4.2.1.

THEOREM 4.4.1. *For an integer $k \ge 0$ assume that $a^{ij}, b^i, c \in C^{k+\delta}(\mathbb{R}^d)$ for any i,j. Then $u \in C^{k+2+\delta}(\mathbb{R}^d)$ whenever $u \in C^{2+\delta}(\mathbb{R}^d)$ and $Lu - zu \in C^{k+\delta}(\mathbb{R}^d)$ for a $z \in C$. Moreover, there is a constant $z_0 > 0$ depending only on d, κ, η, δ and norms of a, b, c in $C^\delta(\mathbb{R}^d)$ such that for any $z \in E_\eta$ with $|z| \ge z_0$ and any $u \in C^{k+2+\delta}(\mathbb{R}^d)$ we have*

$$[u]_{k+2+\delta} + |z|^{(k+2+\delta)/2}|u|_0 \le N\Big([f]_{k+\delta} + |z|^{(k+\delta)/2}|f|_0\Big), \tag{4.4.1}$$

where $f := Lu - zu$ and the constant N depends only on d, κ, η, δ and norms of a, b, c in $C^{k+\delta}(\mathbb{R}^d)$.

To show that this theorem is indeed a particular case of Theorem 4.2.1, we write

$$Lu - zu = a^{ij}u_{x^ix^j} + b^i_z|z|^{1/2}u_{x^i} + c_z|z|u,$$

where $b^i_z = b^i/|z|^{1/2}, c_z = c/|z| - \theta$, $\theta = z/|z|$. Define

$$L^z u := a^{ij}u_{x^ix^j} + b^i_z u_{x^i} + c_z u,$$

and notice that

$$L^z_\lambda u := a^{ij}u_{x^ix^j} + \lambda b^i_z u_{x^i} + \lambda^2 c_z u, \quad L^z_{|z|^{1/2}}u = Lu - zu.$$

The characteristic polynomial of L^z equals $p^z(\xi,x) = -a^{ij}\xi^i\xi^j + \sqrt{-1}\, b^i_z\xi^i + c_z$.

If we formally take $z = \infty$, $|p^z|$ becomes

$$I(\xi,x) := |a^{ij}\xi^i\xi^j + \theta|.$$

When $\arg|\theta| \le \pi/4$,

$$I(\xi,x) \ge |a^{ij}\xi^i\xi^j + 2^{-1/2}| \ge \kappa|\xi|^2 + 2^{-1/2} \ge \kappa_1(1 + |\xi|^2).$$

For $\pi/4 \le \arg|\theta| \le \eta$ we have $\operatorname{Im}\theta \ge (\sin\eta) \wedge \sin(\pi/4)$. Therefore, $I(\xi,x) \ge (\sin\eta) \wedge \sin(\pi/4)$ for any ξ. Also in this case obviously $I(\xi,x) \ge \kappa|\xi|^2/2$ for $|\xi| \ge 2/\kappa$. Therefore, for $\pi/4 \le \arg|\theta| \le \eta$ we also have

$$I(\xi,x) \ge \kappa_2(1 + |\xi|^2) \quad \forall\xi$$

with $\kappa_2 = \kappa_2(\kappa, \eta) > 0$. When $|z|$ is large, $|p^z|$ is close to $I(\xi, x)$. It is seen (cf. Lemma 1.1.7) that when $z \in E_\eta$ and $|z|$ is large enough, say $|z| \geq z_1$, the operator L^z is elliptic and its constant of ellipticity is bounded away from zero because of our assumptions on a, η. For those $|z|$ the norms of b_z, c_z in the relevant Hölder spaces are also uniformly bounded. Therefore, in Theorem 4.1.2 we can take the constant λ_0 to be the same for all L^z with $|z| \geq z_1$ and $z \in E_\eta$. Then it remains only to define $z_0 = z_1 + \lambda_0^2$, apply Theorem 4.2.1 to our L^z_λ, and take $\lambda = \sqrt{|z|}$.

THEOREM 4.4.2. *Let the hypotheses of Theorem 4.4.1 be satisfied and take the constant $z_0 > 0$ from this theorem. Then for any $z \in E_\eta$ with $|z| \geq z_0$, for any $f \in C^{k+\delta}(\mathbb{R}^d)$ there exists a unique solution $u \in C^{k+2+\delta}(\mathbb{R}^d)$ of the equation $Lu(x) - zu(x) = f(x)$, $x \in \mathbb{R}^d$.*

Proof. Theorem 4.4.1 implies that we need only consider the case $k = 0$. In this case one can again apply the method of continuity. Fix a $z \in E_\eta \cap \{|z| \geq z_0\}$ and take any smooth curve $z(t) \in E_\eta \cap \{|z| \geq z_0\}$, $t \in [0, 1]$, such that $z(1) = z$ and $z(0)$ is real. Then for the operators $L - z(t)$ one can almost literally repeat the proof of Theorem 4.3.2 after noticing that by this theorem the equation $Lu - z(0)u = f$ is solvable and the corresponding set T is nonempty. The theorem is proved.

COROLLARY 4.4.3. *If a, b, c, f are infinitely differentiable and their every derivative is bounded, the same holds for the solution u of the equation $Lu - zu = f$.*

This corollary can be localized; see Theorems 7.1.1 and 7.1.2.

REMARK 4.4.4 (cf. Remark 4.3.5). Take an integer $k \geq 0$ and a constant $K > 0$. Theorem 4.4.2 means that if $|a, b, c|_{k+\delta} \leq K$ and we take $z_0 > 0$ from Theorem 4.4.2, then for $z \in E_\eta$ with $|z| \geq z_0$ the operator

$$z - L : C^{k+2+\delta}(\mathbb{R}^d) \to C^{k+\delta}(\mathbb{R}^d)$$

is onto and invertible. Moreover, if we denote by $\mathcal{R}_z$ its inverse operator, then by Theorem 4.4.1 there is a constant N depending only on $\kappa, \eta, \delta, K, d, z_0$ such that for $f \in C^{k+\delta}(\mathbb{R}^d)$

$$[\mathcal{R}_z f]_{k+2+\delta} + |z|^{(k+2+\delta)/2} |\mathcal{R}_z f|_0 \leq N |z|^{(k+\delta)/2} |f|_{k+\delta}.$$

By interpolating we also get that for $r \leq k + 2 + \delta$

$$[\mathcal{R}_z f]_r \leq N z^{r/2-1} |f|_{k+\delta}, \quad |\mathcal{R}_z f|_r \leq N z^{r/2-1} |f|_{k+\delta}.$$

In particular, we have

$$|\mathcal{R}_z f|_2 \leq N |f|_\delta, \quad |\mathcal{R}_z f|_1 \leq |z|^{-1/2} N |f|_\delta,$$

$$|\mathcal{R}_z f|_{1+\delta} \leq |z|^{(\delta-1)/2} N |f|_\delta, \quad |\mathcal{R}_z f|_\delta \leq |z|^{-1+\delta/2} N |f|_\delta.$$

Theorem 4.4.2 also implies that the operators $\mathcal{R}_z$ are independent of k and δ.

EXERCISE* 4.4.5. Take z as in Theorem 4.4.2, and take the operators $\mathcal{R}_z$ from Remark 4.4.4. Then we have a function $\mathcal{R}_z$ with values in the Banach space of linear bounded operators mapping $C^{\delta}(\mathbb{R}^d)$ into $C^{2+\delta}(\mathbb{R}^d)$. Prove that the operator function $z \to \mathcal{R}_z$, which is called the *resolvent* of L, is analytic and $|\mathcal{R}_z|_{C^{\delta}\to C^{\delta}} \leq N|z|^{-1+\delta/2}$ for z under consideration. Conclude, in particular, that for any $f \in C^{\delta}(\mathbb{R}^d)$ and $x \in \mathbb{R}^d$ the function $\mathcal{R}_z f(x)$ is an analytic function of z.

EXERCISE 4.4.6. Prove that $\mathcal{R}_z$ is well defined if $\operatorname{Re} z > 0$, and $|\mathcal{R}_z f|_0 \leq (\operatorname{Re} z)^{-1}|f|_0$.

4.5. Solvability of higher–order elliptic equations with variable coefficients

THEOREM 4.5.1. *Let $L = L(x) = \sum_{|\alpha|\leq m} a^{\alpha}(x)D^{\alpha}$ be a uniformly elliptic operator, and $k \geq 0$ be an integer. Assume that $a^{\alpha} \in C^{k+\delta}(\mathbb{R}^d)$ for any α. Define the constant λ_0 depending only on the ellipticity constant κ and m, δ, d and maximum of $|a^{\alpha}|_{\delta}$ as in Theorem 4.1.2, and take any real λ such that $|\lambda| \geq \lambda_0$. Then for any $f \in C^{k+\delta}(\mathbb{R}^d)$ there exists a unique solution $u \in C^{k+m+\delta}(\mathbb{R}^d)$ of the equation $L_{\lambda}u(x) = f(x)$, $x \in \mathbb{R}^d$.*

Proof. The uniqueness follows at once from Theorem 4.1.2. From Theorem 4.2.1 we also infer that we need to consider only the case $k = 0$. The proof of the existence we break into four steps. We will again apply the method of continuity. Notice that if L is a second–order elliptic operator and for some x we have $L = \Delta - 1$ and for other x we have $L = -\Delta + 1$, the "naive" way of joining L with $\Delta - 1$ as $tL + (1-t)(\Delta - 1)$ does not produce a family of uniformly elliptic operators. Therefore we use another family of operators.

Step 1. Assume that a^{α} are independent of x. Then we get the result from Theorem 3.7.2.

Step 2. Assume that $a^{\alpha} \in C^1(\mathbb{R}^d)$. For $t \in [0, \infty]$ (including ∞) define

$$\zeta(t,x) = \frac{tx}{t + |x|}, \quad a^{\alpha}(t,x) = a^{\alpha}(\zeta(t,x)), \quad L_{\lambda}^t = \sum_{|\alpha|\leq m} a^{\alpha}(t,x)\lambda^{m-|\alpha|}D^{\alpha}$$

(of course, $\zeta(\infty, x) := x$) and let T be the set of all points $t \in [0,\infty]$ for which the statement of the theorem is true (with L_{λ}^t in place of L_{λ}). Observe that $a^{\alpha}(0,x) = a^{\alpha}(0)$. Therefore, the set T is not empty by the above.

Next, we claim that

$$|\zeta(t,x) - \zeta(t,y)| \leq |x - y| \tag{4.5.1}$$

for any t, x, y. Obviously we need only to show that $|\operatorname{grad}_x \zeta(t,x)| \leq 1$ for $x \neq 0$. For $x \neq 0$ and unit l we have

$$|l \cdot \operatorname{grad}_x \zeta(t,x)| = \left|\frac{tl}{t+|x|} - \frac{tx}{(t+|x|)^2}\frac{l\cdot x}{|x|}\right|$$

$$\leq \frac{t}{t+|x|} + \frac{t|x|}{(t+|x|)^2} \leq \frac{t}{t+|x|} + \frac{|x|}{t+|x|} = 1.$$

This proves (4.5.1).

Inequality (4.5.1) shows, in particular, that $[a^\alpha(t,\cdot)]_\delta \le [a^\alpha]_\delta$, so that by Theorem 4.1.2 inequality (3.7.3) holds with the same constant N for all solutions of the equations $L^t_\lambda u = f$. This immediately implies that T is closed.

Therefore, to finish this step it remains only to prove that $T \cap [0,\infty)$ is open in the topology of $[0,\infty)$. Take any point $t_0 \in T \cap [0,\infty)$ and define the linear operator $\mathcal{R} : C^{m+\delta}(\mathbb{R}^d) \to C^\delta(\mathbb{R}^d)$ such that it takes any $f \in C^\delta(\mathbb{R}^d)$ into the solution of $L^{t_0}_\lambda u = f$. By the assumption $\mathcal{R}$ is well defined and by Theorem 4.1.2 is bounded. Now in order to show that for $t \in [0,\infty)$ close to t_0 the equation $L^t_\lambda u = f$ is solvable we write this equation as

$$L^{t_0}_\lambda u = f + (L^{t_0}_\lambda - L^t_\lambda)u, \quad u = \mathcal{R}f + \mathcal{R}(L^{t_0}_\lambda - L^t_\lambda)u,$$

and we show that the operator $\mathcal{R}(L^{t_0}_\lambda - L^t_\lambda)$ is a contraction in $C^{m+\delta}(\mathbb{R}^d)$ for $t \in [0,\infty)$ close to t_0.

By the above

$$|\mathcal{R}(L^{t_0}_\lambda - L^t_\lambda)u|_{m+\delta} \le N|(L^{t_0}_\lambda - L^t_\lambda)u|_\delta$$

with the constant N independent of t, u. Next

$$|(L^{t_0}_\lambda - L^t_\lambda)u|_\delta \le \sum_{|\alpha|\le m} \{[a^\alpha(t_0,\cdot) - a^\alpha(t,\cdot)]_\delta |D^\alpha u|_0 + |a^\alpha(t_0,\cdot) - a^\alpha(t,\cdot)|_0 |D^\alpha u|_\delta\}.$$

Here

$$|a^\alpha(t_0,\cdot) - a^\alpha(t,\cdot)|_0 \le N \sup_x |\zeta(t_0,x) - \zeta(t,x)| \le N|t_0 - t|$$

since $|\zeta_t| \le 1$. Also if $|x-y| \le |t_0 - t|$, then (by smoothness of a^α)

$$I := \frac{|[a^\alpha(t_0,x) - a^\alpha(t,x)] - [a^\alpha(t_0,y) - a^\alpha(t,y)]|}{|x-y|^\delta} \le \frac{|a^\alpha(t_0,x) - a^\alpha(t_0,y)|}{|x-y|^\delta}$$

$$+ \frac{|a^\alpha(t,x) - a^\alpha(t,y)|}{|x-y|^\delta} \le N|x-y|^{1-\delta} \le N|t_0 - t|^{1-\delta}.$$

If $|x-y| \ge |t_0 - t|$, then

$$I \le \frac{|a^\alpha(t_0,x) - a^\alpha(t,x)|}{|x-y|^\delta} + \frac{|a^\alpha(t_0,y) - a^\alpha(t,y)|}{|x-y|^\delta}$$

$$\le N\frac{|t_0 - t|}{|x-y|^\delta} \le N|t_0 - t|^{1-\delta}.$$

Thus, for $|t_0 - t| \le 1$ we have

$$|(L^{t_0}_\lambda - L^t_\lambda)u|_\delta \le N|u|_{m+\delta}|t_0 - t|^{1-\delta}, \quad |\mathcal{R}(L^{t_0}_\lambda - L^t_\lambda)u|_{m+\delta} \le N|u|_{m+\delta}|t_0 - t|^{1-\delta}.$$

Since the last constant N is independent of t, u, we can indeed assert that the operator $\mathcal{R}(L^{t_0}_\lambda - L^t_\lambda)$ is a contraction for t close to t_0, and this finishes Step 2.

Step 3. We now pass to the general situation and prove the solvability if $|\lambda|$ is large enough. For any a^α construct the functions $a^{\varepsilon\alpha}$ as in Exercise 3.2.3. Observe that by this Exercise $[a^{\varepsilon\alpha}]_\delta \le [a^\alpha]_\delta$. Also $a^{\varepsilon\alpha}$ are uniformly close to a^α if ε is small since $a^\alpha \in C^\delta(\mathbb{R}^d)$. Therefore, if ε is small, then for the operators

$L^\varepsilon = \sum_{|\alpha|\le m} a^{\varepsilon\alpha} D^\alpha$ the ellipticity constant κ^ε is bigger than, say, one half of the ellipticity constant κ of L (see Lemma 1.1.7) and the quantities λ_0^ε corresponding to L^ε are uniformly bounded, say, by $\bar\lambda_0$. Finally, $a^{\varepsilon\alpha} \in C_b^\infty(\mathbb{R}^d)$. By Step 2 we can solve the equations $L_\lambda^\varepsilon u_\varepsilon = f$ whenever $|\lambda| \ge \bar\lambda_0$, $f \in C^\delta(\mathbb{R}^d)$. Moreover, the norms of u_ε in $C^{m+\delta}(\mathbb{R}^d)$ are bounded independently of ε. Therefore we can again use the familiar passage to the limit.

Step 4. We leave to the reader to prove that by the method of continuity, solvability of $L_\lambda u = f$ for all λ with $|\lambda| \ge \lambda_0$ (for which we have the a priori estimate) can be derived from solvability for large $|\lambda|$. The theorem is proved.

EXERCISE* 4.5.2. Complete Step 4 in the above proof.

COROLLARY 4.5.3. *If a^α are infinitely differentiable and their every derivative is bounded, the same holds for the solution u of the equation $L_\lambda u = f$.*

This corollary can be localized; see Theorems 7.1.1 and 7.1.2.

4.6. Hints to exercises

4.1.4. Observe that $[\Delta u - u]_\delta \le [\Delta u + F(\cdot, u_x)]_\delta + [F(\cdot, u_x) + u]_\delta$, and use Exercise 3.2.7 and estimate $[u]_2$ through $[u]_{2+\delta}$ and $|u|_0$ to show that $[F(\cdot, u_x) + u]_\delta \le \varepsilon|u|_{2+\delta} + N(\varepsilon, |u|_0)$. In particular apply *Young's inequality*: $ab \le p^{-1}(\varepsilon a)^p + q^{-1}(b/\varepsilon)^q$ if $p, q > 1$, $p^{-1} + q^{-1} = 1$.

4.2.3. Use Theorem 4.2.1 with $\lambda = \lambda_0$ and as in its proof observe that $L_{\lambda_0} - L_0$ as well as $L_{\lambda_0} - L$ is an operator of order less than or equal to $m - 1$.

4.2.4. Observe that

$$0 = \delta_{h,j} F(x, D^\alpha u(x)) = \sum_{|\alpha|\le m} a^\alpha_{h,j}(x) D^\alpha(\delta_{h,j}u)(x) + f_{h,j}(x),$$

where

$$a^\beta_{h,j}(x) = \int_0^1 F_{\xi(\beta)}(x + the_j, tD^\alpha u(x + he_j) + (1-t)D^\alpha u(x))\, dt,$$

$$f_{h,j}(x) = \int_0^1 F_{x^j}(x + the_j, tD^\alpha u(x + he_j) + (1-t)D^\alpha u(x))\, dt.$$

By using Exercises 4.2.3 and 3.2.8 first prove a $C^{m+\delta^2}(\mathbb{R}^d)$-estimate of $\delta_{h,j}u$ independent of h if h is small enough. Conclude that $u \in C^{1+m+\delta^2}(\mathbb{R}^d)$, in particular, $u \in C^{1+m}(\mathbb{R}^d)$, which along with Exercise 3.2.8 shows that $a^\alpha_{h,j}$ and $f_{h,j}$ are C^δ-functions. Then repeat the previous argument. If $k \ge 2$, one can differentiate the equation $F(x, D^\alpha u) = 0$ and use a bootstrap argument.

4.3.6. Observe that with some bounded b we have $f = \Delta u + b^i u_{x^i} - u$ and use Theorem 2.9.2.

4.3.7. First estimate $|u|_0$ almost as in Exercise 4.3.6 (although we do not assume that $F(0) = 0$). Then $|u_x|_{2+\delta}$ remains to be estimated. Further, observe that for any $i = 1, ..., d$ the function $v = u_{x^i}$ satisfies $\Delta v - b^i v_{x^i} - v = f_{x^i}$ where $b^i = F_{\xi^i}(\operatorname{grad} u)$. Conclude that $|u_x|_0 \le |f_x|_0$, so that $|f - F(\operatorname{grad} u)|_0$ is under control. This and Exercise 3.4.5 lead to an estimate of $|u|_{1+\delta}$, which estimates Hölder's constant of b^i.

4.3.9. Use Exercises 4.3.6 and 4.3.8 to prove an a priori estimate and uniqueness and the fact that one may assume that F is twice continuously differentiable. After this *fix* $f \in C^\delta(\mathbb{R}^d)$ and define T as the set of all $t \in [0,1]$ such that the equation $\Delta u + tF(\operatorname{grad} u) - u = f$ has a unique solution $u \in C^{2+\delta}(\mathbb{R}^d)$. In the proof that T is open take a point $t_0 \in T$ and the corresponding solution u_0 and define $b^i(x) = F_{u^i}(\operatorname{grad} u_0(x))$. Also denote by $\mathcal{R}_t f$ the mapping which brings $w \in C^{2+\delta}(\mathbb{R}^d)$ into a unique solution $v \in C^{2+\delta}(\mathbb{R}^d)$ of the linear equation

$$\Delta v + tb^i v_{x^i} - v = f - t\{F(\operatorname{grad} w) - b^i w_{x^i}\}.$$

Show that there exist $\varepsilon > 0$ and $\gamma > 0$ such that if $w \in \mathcal{B}_\varepsilon := \{u \in C^{2+\delta}(\mathbb{R}^d) : |u - u_0|_{2+\delta} \le \varepsilon\}$ and $|t_0 - t| \le \gamma$, then $v \in \mathcal{B}_\varepsilon$, so that $\mathcal{R}_t : \mathcal{B}_\varepsilon \to \mathcal{B}_\varepsilon$ and, moreover, the mapping $\mathcal{R}_t : \mathcal{B}_\varepsilon \to \mathcal{B}_\varepsilon$ is a contraction.

4.3.11. Make necessary changes in the hint to Exercise 4.3.9. Alternately, you may just use the results of Exercises 4.3.7 and 4.3.9.

4.4.5. Remember that we say that a Banach space–valued function of the complex variable z is analytic in a domain if in a neighborhood of any point it admits a power series representation. From the definition of $\mathcal{R}_z$ deduce that $\mathcal{R}_{z_1} = \mathcal{R}_{z_2} + (z_2 - z_1)\mathcal{R}_{z_2}\mathcal{R}_{z_1}$, then iterate this formula. The last formula becomes natural if one looks at $\mathcal{R}_z$ as $(z - L)^{-1}$. In its proof you may need to prove that $\mathcal{R}_{z_1}$ and $z_2 - L$ commute on sufficiently regular functions.

4.4.6. First reduce the problem to proving that $|u|_0 \le (\operatorname{Re} z)^{-1}|zu - Lu|_0$ for any $u \in C^{2+\delta}(\mathbb{R}^d)$. Then remember that for real z we have this inequality, and take z_1 real and observe that $|Lu - z_1 u|_0 \le |Lu - zu|_0 + |z_1 - z| \cdot |u|_0$, where $z_1 - |z_1 - z|$ is close to $\operatorname{Re} z$ if z_1 is large enough.

CHAPTER 5

Second–Order Elliptic Equations in Half Spaces

The simplest case of equations in domains is given by equations in a half space. Their investigation along with the investigation of equations in the whole space will play a crucial role in our treatment of equations in bounded domains.

If we are given an elliptic equation $Lu = f$ in a domain Ω, then in order to define the function u uniquely we also have to impose some boundary conditions on u. The number of conditions depends on the order of the equation; it is equal to one only for second–order equations. The case of second–order equations is therefore the simplest one, and we consider only this case below. More precisely we will be dealing only with the Dirichlet problem for second–order elliptic equations, confining ourselves to mere comments about the Neumann and oblique derivative problems which involve first–order derivatives. We also comment on second–order boundary conditions.

Furthermore, in the next chapter we want to consider second–order equations $Lu = f$ with homogeneous elliptic operators L, of which Laplace's operator Δ is an example. However, if the operator L has complex coefficients, some unpleasant things may occur. For example, for $d = 2$ for the Bitsadze equation

$$\frac{\partial^2}{(\partial \bar{z})^2} u = 0$$

with the homogeneous elliptic operator

$$\frac{\partial^2}{(\partial \bar{z})^2} = \frac{\partial^2}{(\partial x)^2} + 2i \frac{\partial^2}{\partial x \partial y} - \frac{\partial^2}{(\partial y)^2}$$

any function of the form

$$u(x, y) = f(z)(1 - |z|^2) = f(z)(1 - z\bar{z})$$

with analytic f is a solution which equals zero on the boundary of the unit disk (remember that $(\partial/\partial\bar{z})f(z) = 0$ for any analytic function). In this situation we lose uniqueness. Therefore, we will consider only elliptic equations with *real–valued coefficients*, which have been introduced in Section 4.4. For the sake of simplicity we will not deal with complex parameters z, although almost everything below could be repeated without change for complex z as well and could produce considerable information about the spectrum of L. Finally, all functions $u, v, \ldots$ will be assumed to be real–valued unless stated otherwise explicitly.

As always the reader aiming at $C^{2+\delta}$–theory rather than $C^{k+2+\delta}$–theory should take $k = 0$ everywhere in this and subsequent chapters.

5.1. More equivalent norms in the Hölder spaces

Denote

$$\mathbb{R}^d_+ = \{x = (x', x^d) : x' = (x^1, ..., x^{d-1}) \in \mathbb{R}^{d-1}, x^d > 0\},$$

$$\partial_{h,j} u(x) = u(x + he_j) - u(x), \quad \partial_h^\alpha = \partial_{h,1}^{\alpha_1} \cdot ... \cdot \partial_{h,d}^{\alpha_d}.$$

Observe that since $h > 0$, the quantities $\partial_h^\alpha u(x)$ are well defined for any function u defined in $\mathbb{R}^d_+$. We will also use the fact that the operators ∂_h^α commute.

To understand better why we need Theorem 5.1.3, do the following exercise.

EXERCISE 5.1.1. By using Exercise 2.6.11 show that if $\Omega = \mathbb{R}^d_+$, $\delta \in (0,1)$ and $u \in C^{2+\delta}(\bar{\Omega})$ and $\Delta u = 0$ in Ω, then for $N = N(d, \delta)$ and any $i, j = 1, ..., d-1$ we have

$$|u_{x^i x^j}|_{\delta;\Omega} \leq N |u_{x^i x^j}|_{\delta;\partial\Omega}, \quad |u_{x^d x^d}|_{\delta;\Omega} \leq N \sum_{i=1}^{d-1} |u_{x^i x^i}|_{\delta;\partial\Omega}.$$

LEMMA 5.1.2. *Let Ω be $\mathbb{R}^d$ or $\mathbb{R}^d_+$, and let $k \geq 0$ be an integer. For a function u defined in Ω and $0 < \delta < 1$ denote*

$$[u]'_{k+\delta;\Omega} = \sup_{|\alpha|=k+1} \sup_{h>0} \sup_{x\in\Omega} \frac{1}{h^{k+\delta}} |\partial_h^\alpha u(x)|.$$

Then there is a constant $N = N(d, k, \delta)$ such that for any $u \in C^{k+\delta}(\Omega)$ we have

$$[u]_{k+\delta;\Omega} \leq N [u]'_{k+\delta;\Omega}, \quad [u]'_{k+\delta;\Omega} \leq N [u]_{k+\delta;\Omega}. \tag{5.1.1}$$

Proof. Take $h > 0$ and α such that $|\alpha| = k+1$. Define j, β so that $\alpha = \beta + e_j$ with $|\beta| = k$. By the mean value theorem for any $x \in \Omega$ there is a $y \in \Omega$ such that $|x - y| \leq kh$ and

$$\partial_h^\alpha u(x) = \partial_h^\beta \partial_{h,j} u(x) = h^k D^\beta \Big(\partial_{h,j} u\Big)(y) = h^k \partial_{h,j} D^\beta u(y).$$

Hence,

$$|\partial_h^\alpha u(x)| \leq h^k |D^\beta u(y + he_j) - D^\beta u(y)| \leq h^{k+\delta} [u]_{k+\delta;\Omega}.$$

This proves the second inequality in (5.1.1).

In order to prove the first one, we fix an integer n which will be specified later and take $|\beta| = k$. After breaking the straight–line interval between $x + he_j$ and x into n pieces we see that

$$|D^\beta u(x + he_j) - D^\beta u(x)| \leq |D^\beta u(x + he_j) - n^k h^{-k} \partial_{h/n}^\beta u(x + he_j)|$$

$$+ n^k h^{-k} \sum_{r=0}^{n-1} |\partial_{h/n,j} \partial_{h/n}^\beta u(x + rhe_j/n)| + |n^k h^{-k} \partial_{h/n}^\beta u(x) - D^\beta u(x)|.$$

As follows from the above, the first and the last terms on the right are less than $N(h/n)^\delta [u]_{k+\delta;\Omega}$, where the constant N depends only on d, k. Also by definition the terms in the sum with respect to r are less than $[u]'_{k+\delta;\Omega}(h/n)^{k+\delta}$. Therefore,

$$\sup_{x\in\Omega,h>0}\frac{|D^\beta u(x+he_j)-D^\beta u(x)|}{h^\delta}\leq Nn^{-\delta}[u]_{k+\delta;\Omega}+n^{1-\delta}[u]'_{k+\delta;\Omega}.$$

Since j is arbitrary, this certainly implies that

$$[u]_{k+\delta;\Omega}\leq Nn^{-\delta}[u]_{k+\delta;\Omega}+n^{1-\delta}[u]'_{k+\delta;\Omega},$$

and upon choosing n so that $Nn^{-\delta}\leq 1/2$, we conclude the proof. The lemma is proved.

The next theorem is to be compared with Exercise 3.7.3 (also see Remark 5.3.4).

THEOREM 5.1.3. *Let Ω be $\mathbb{R}^d$ or $\mathbb{R}^d_+$ and take $u\in C^{2+\delta}(\Omega)$. Then*

$$[u]_{2+\delta;\Omega}\leq N(d,\delta)\max\{[u_{x^ix^j}]_{\delta;\Omega},[u_{x^dx^d}]_{\delta;\Omega}:i,j=1,...,d-1\}.$$

Proof. For any fixed $i,j,k=1,...,d$ at least two of i,j,k either coincide with d or are less than or equal to $d-1$. In the latter case assume that $i,j\leq d-1$. Then for any $h>0$ and $x\in\Omega$ there is a $y\in\Omega$ such that

$$|\partial_{h,i}\partial_{h,j}\partial_{h,k}u(x)|=h^2|\partial_{h,k}u_{x^ix^j}(y)|\leq h^{2+\delta}[u_{x^ix^j}]_\delta.$$

In the remaining case if, say, $i=j=d$, then for any $h>0$ and $x\in\Omega$ there is a $y\in\Omega$ such that

$$|\partial_{h,i}\partial_{h,j}\partial_{h,k}u(x)|=|\partial_{h,k}\partial^2_{h,d}u(x)|=h^2|\partial_{h,k}u_{x^dx^d}(y)|\leq h^{2+\delta}[u_{x^dx^d}]_\delta.$$

It remains only to apply Lemma 5.1.2. The theorem is proved.

5.2. Laplace's equation in half spaces

In this section we assume that $d\geq 2$ and, as always, that $\delta\in(0,1)$.

Exercise 5.1.1 and Theorem 5.1.3 provide us with the basic a priori estimate (5.2.2) for half spaces. Nevertheless, we prove it differently here because, first, Exercise 5.1.1 is left to the reader and, second, we need not only a priori estimates but also existence theorems.

LEMMA 5.2.1. *For $x\in\mathbb{R}^d_+, y'\in\mathbb{R}^{d-1}$ define the Poisson kernel*

$$P_d(x,y')=b_d\frac{x^d}{(|x'-y'|^2+|x^d|^2)^{d/2}},\quad b_d^{-1}:=\int_{\mathbb{R}^{d-1}}\frac{1}{(|y'|^2+1)^{d/2}}\,dy'.$$

Then for any bounded continuous function $g(y')$ on $\mathbb{R}^{d-1}$ the function $u(x)$, $x^d>0$, defined by

$$\begin{aligned}u(x):=\Pi g(x):=&\int_{\mathbb{R}^{d-1}}P_d(x,y')g(y')\,dy'\\ =&b_d\int_{\mathbb{R}^{d-1}}\frac{1}{(|y'|^2+1)^{d/2}}g(x^dy'+x')\,dy'\end{aligned}\tag{5.2.1}$$

has a continuous extension to $\bar{\mathbb{R}}^d_+$ which is equal to g on $\mathbb{R}^{d-1}$. Furthermore, u is infinitely differentiable in $\mathbb{R}^d_+$ and is a harmonic function in $\mathbb{R}^d_+$. Finally, if $g\in C^{2+\delta}(\mathbb{R}^{d-1})$, then $u\in C^{2+\delta}(\mathbb{R}^d_+)$ and

$$[u]_{2+\delta;\mathbb{R}^d_+} \le N[g]_{2+\delta;\mathbb{R}^{d-1}}, \tag{5.2.2}$$

where the constant N depends only on d and δ.

Proof. Observe that the second equality in (5.2.1) follows after the change of variables $y' \to x^d y' + x'$, and this formula along with the dominated convergence theorem proves the first assertion of the lemma. By using the first equality one proves the smoothness of u and the fact that $\Delta u = 0$ in $\mathbb{R}^d_+$ in a straightforward way starting from the fact that $\Delta_x P_d(x, y') = 0$.

To prove (5.2.2) we first take derivatives with respect to the tangential variables x' and we observe that for $i, j = 1, 2, ..., d-1$ (cf. Exercise 5.1.1)

$$|u_{x^i x^j}(x_1) - u_{x^i x^j}(x_2)|$$

$$\le b_d \int_{\mathbb{R}^{d-1}} \frac{1}{(|y'|^2+1)^{d/2}} |g_{x^i x^j}(x_1^d y' + x_1') - g_{x^i x^j}(x_2^d y' + x_2')|\, dy'$$

$$\le b_d [g]_{2+\delta;\mathbb{R}^{d-1}} |x_1 - x_2|^\delta \int_{\mathbb{R}^{d-1}} \frac{(|y'|^2+1)^{\delta/2}}{(|y'|^2+1)^{d/2}}\, dy' = N[g]_{2+\delta;\mathbb{R}^{d-1}} |x_1 - x_2|^\delta.$$

From the equation $\Delta u = 0$ we can express $u_{x^d x^d}$ through $u_{x^j x^j}$ with $j \le d-1$, and this proves that $[u_{x^d x^d}]_{\delta;\mathbb{R}^d+} \le N[g]_{2+\delta;\mathbb{R}^{d-1}}$.

Finally, application of Theorem 5.1.3 brings the proof to an end. Of course, we should add that first we apply this theorem to the domain $\varepsilon e_d + \mathbb{R}^d_+$ in which u has all derivatives bounded, so that it really belongs to the class $C^{2+\delta}$ in this domain, and then because of the arbitrariness of ε we get the estimate for $\mathbb{R}^d_+$. The lemma is proved.

REMARK 5.2.2. One of the ways to find the kernel P_d and formula (5.2.1) is to solve the equation $\Delta u = 0$ in $\mathbb{R}^d_+$ with the boundary condition $u = g$ on $\mathbb{R}^{d-1}$. To solve this problem one can apply the Fourier transform with respect to the tangential variables. Then in obvious notation one gets $-|\xi'|^2 \tilde{u}(\xi', x^d) + \tilde{u}_{x^d x^d}(\xi', x^d) = 0$ and $\tilde{u}(\xi', 0) = \tilde{g}(\xi')$. One solves this for any $\xi' \in \mathbb{R}^{d-1}$ to find $\tilde{u}(\xi', x^d) = \tilde{g}(\xi') \exp(-|\xi'| x^d)$, and then one applies the inverse Fourier transform. Another way is described in Exercise 2.3.3.

LEMMA 5.2.3. *For a bounded continuous function $g(y')$ on $\mathbb{R}^{d-1}$ and $x \in \bar{\mathbb{R}}^d_+$ define*

$$v(x) := \Pi_1 g(x) := b_{d+1} \int_{\mathbb{R}^{d-1}} \int_{-\infty}^{\infty} \frac{1}{(|y'|^2 + t^2 + 1)^{(d+1)/2}} e^{itx^d} g(x^d y' + x')\, dy' dt.$$

Then the function v is continuous in $\bar{\mathbb{R}}^d_+$, $v(x', 0) = g(x')$ on $\mathbb{R}^{d-1}$. Moreover, v is infinitely defferentiable in $\mathbb{R}^d_+$ and $\Delta v - v = 0$ in $\mathbb{R}^d_+$. Finally, if $g \in C^{2+\delta}(\mathbb{R}^{d-1})$, then $v \in C^{2+\delta}(\mathbb{R}^d_+)$ and

$$[v]_{2+\delta;\mathbb{R}^d_+} \le N|g|_{2+\delta;\mathbb{R}^{d-1}},$$

where the constant N depends only on d and δ.

Proof. If one adds one more independent variable $t \in (-\infty, \infty)$ and defines $g(t, x') = g(x') \exp(it)$, then one easily sees that

$$v(x)e^{is} = \int_{\mathbb{R}^d} P_{d+1}((s,x),(t,y'))g(t,y')\,dtdy'.$$

Therefore, all our assertions follow directly from Lemma 5.2.1. The lemma is proved.

EXERCISE 5.2.4. The restriction $0 < \delta < 1$ is essential for the last assertions in Lemma 5.2.1 or Lemma 5.2.3. To show this for $d = 2$ and $\mathbb{R}^2 = \{(x, y) : x, y \in \mathbb{R}\}$ take a bounded function $g(x)$ to be $|x|$ for x close to zero and to satisfy a Lipschitz condition on $(-\infty, \infty)$. Take $u(x, y)$ from Lemma 5.2.1 or Lemma 5.2.3 and prove that u is *not* Lipschitz continuous in y.

5.3. Poisson's equation in half spaces

Denote

$$C^r_0(\bar{\mathbb{R}}^d_+) = C^r(\bar{\mathbb{R}}^d_+) \cap \{u : u(x',0) \equiv 0\}.$$

THEOREM 5.3.1. *For any $f \in C^\delta(\mathbb{R}^d_+)$ there exists a unique function $u \in C^{2+\delta}_0(\bar{\mathbb{R}}^d_+)$ such that $\Delta u - u = f$ in $\mathbb{R}^d_+$. Moreover,*

$$|u|_{2+\delta;\mathbb{R}^d_+} \le N|f|_{\delta;\mathbb{R}^d_+}, \tag{5.3.1}$$

where the constant N is independent of f.

Proof. It is natural to continue f on $\mathbb{R}^d$ to get an odd function and then solve the equation $\Delta u - u = f$ in $\mathbb{R}^d$. Because of uniqueness this solution should vanish on $\mathbb{R}^{d-1}$, and its restriction on $\mathbb{R}^d_+$ should be the function we need. The only trouble is that the odd continuation of f need not be continuous. That is why we slightly modify this idea.

First we define a function $g = g(x')$ as a unique solution in $C^{2+\delta}(\mathbb{R}^{d-1})$ of the equation

$$\Delta_{d-1}g(x') - g(x') = f(x',0),$$

where Δ_{d-1} denotes Laplace's operator with respect to $x^1, ..., x^{d-1}$. Then, we take the function v from Lemma 5.2.3. By Theorems 3.7.1 and 3.7.2 and Lemma 5.2.3 for the functions g, v we have

$$|g|_{2+\delta;\mathbb{R}^{d-1}} \le N|f|_{\delta;\mathbb{R}^d_+}, \quad [v]_{2+\delta;\mathbb{R}^d_+} \le N|g|_{2+\delta;\mathbb{R}^{d-1}} \le N|f|_{\delta;\mathbb{R}^d_+}. \tag{5.3.2}$$

Next, let $\bar{f}(x) = \operatorname{sign}(x^d)[f(x',|x^d|) - f(x',0)]$, so that f is odd with respect to x^d. Observe that $|\bar{f}|_{\delta;\mathbb{R}^d} \le 2|f|_{\delta;\mathbb{R}^d_+}$ since $\bar{f} = 0$ on the border of $\mathbb{R}^d_+$, and there exists a unique solution $\bar{u} \in C^{2+\delta}(\mathbb{R}^d)$ of the equation $\Delta\bar{u} - \bar{u} = \bar{f}$. It follows from uniqueness that $\bar{u}$ is odd with respect to x^d, so that $\bar{u}(x',0) \equiv 0$. Also

$$[\bar{u}]_{2+\delta;\mathbb{R}^d_+} \le N|\bar{f}|_{\delta;\mathbb{R}^d} \le N|f|_{\delta;\mathbb{R}^d_+}. \tag{5.3.3}$$

Finally for the function $u(x) := \bar{u}(x) + g(x') - v(x)$, $x^d \ge 0$, we have $\Delta u - u = f$ in $\mathbb{R}^d_+$ and $u(x',0) \equiv 0$. To get (5.3.1) it remains only to collect (5.3.2) and (5.3.3) and add that by Theorem 2.9.2 $|u|_{0;\mathbb{R}^d_+} \le |f|_{0;\mathbb{R}^d_+}$. The uniqueness of u we know even under more general circumstances (see Corollary 2.9.3). The theorem is proved.

Next we extend Theorem 5.3.1 to smoother functions f.

THEOREM 5.3.2. *Fix an integer $k \geq 0$. Then for any $f \in C^{k+\delta}(\mathbb{R}^d_+)$ and $g \in C^{2+k+\delta}(\mathbb{R}^{d-1})$ there exists a unique function $u \in C^{k+2+\delta}(\bar{\mathbb{R}}^d_+)$ such that $\Delta u - u = f$ in $\mathbb{R}^d_+$ and $u(x', 0) \equiv g(x')$. Moreover,*

$$|u|_{k+2+\delta;\mathbb{R}^d_+} \leq N\Big(|f|_{k+\delta;\mathbb{R}^d_+} + |g|_{k+2+\delta;\mathbb{R}^{d-1}}\Big), \tag{5.3.4}$$

where the constant N is independent of f, g.

Proof. By writing down the equation for $u(x) - g(x')$ instead of u we reduce the problem to the case $g \equiv 0$, the only one to be considered. The existence and uniqueness of a function $u \in C^{2+\delta}_0(\mathbb{R}^d_+)$ solving the equation $\Delta u - u = f$ is known from Theorem 5.3.1. The only things we have to show are that $u \in C^{k+2+\delta}_0(\bar{\mathbb{R}}^d_+)$ and that (5.3.4) holds.

We will use induction on k. Assume that for $k = k_0$ the theorem is true, and let $f \in C^{k_0+1+\delta}(\mathbb{R}^d_+)$. For $j = 1, ..., d-1$ and $h > 0$ we have in $\mathbb{R}^d_+$ that

$$\Delta(\delta_{h,j} u) - \delta_{h,j} u = \delta_{h,j} f.$$

Hence

$$[\delta_{h,j} u]_{k_0+2+\delta;\mathbb{R}^d_+} \leq N|\delta_{h,j} f|_{k_0+\delta;\mathbb{R}^d_+} \leq N|f|_{k_0+1+\delta;\mathbb{R}^d_+}.$$

This implies that $D_j u \in C^{k_0+2+\delta}_0(\bar{\mathbb{R}}^d_+)$ and

$$[D_j u]_{k_0+2+\delta;\mathbb{R}^d_+} \leq N|f|_{k_0+1+\delta;\mathbb{R}^d_+}, \quad [D_j D_i u]_{k_0+1+\delta;\mathbb{R}^d_+} \leq N|f|_{k_0+1+\delta;\mathbb{R}^d_+}$$

for any $i = 1, ..., d$. Finally,

$$D_d^2 u = f + u - \sum_{j=1}^{d-1} D_j^2 u \in C^{k_0+1+\delta}(\mathbb{R}^d_+),$$

$$[D_d^2 u]_{k_0+1+\delta;\mathbb{R}^d_+} \leq N|f|_{k_0+1+\delta;\mathbb{R}^d_+} + [u]_{k_0+1+\delta;\mathbb{R}^d_+} \leq N|f|_{k_0+1+\delta;\mathbb{R}^d_+}.$$

This proves the theorem.

COROLLARY 5.3.3. *There is a constant $N = N(d, k)$ such that for any $u \in C^{k+2+\delta}(\bar{\mathbb{R}}^d_+)$ we have*

$$[u]_{k+2+\delta;\mathbb{R}^d_+} \leq N([\Delta u]_{k+\delta;\mathbb{R}^d_+} + [u(\cdot, 0)]_{k+2+\delta;\mathbb{R}^{d-1}}). \tag{5.3.5}$$

Indeed by Theorem 5.3.2 for $v(x) = u(x) - u(x', 0)$ we have

$$[v]_{k+2+\delta;\mathbb{R}^d_+} \leq N|\Delta v - v|_{k+\delta;\mathbb{R}^d_+}.$$

For any $\lambda > 0$, as before, this implies

$$[v]_{k+2+\delta;\mathbb{R}^d_+} \leq N\Big([\Delta v - \lambda^2 v]_{k+\delta;\mathbb{R}^d_+} + \sum_{r=0}^{k} \lambda^{k-r+\delta}|\Delta v - \lambda^2 v|_{r;\mathbb{R}^d_+}\Big)$$

with the same constant N as in (5.3.4), and by letting $\lambda \to 0$ we obtain (5.3.5) for v instead of u. Then we use $[u] \leq [v] + [u(\cdot, 0)]$, $[\Delta v] \leq [\Delta u] + [\Delta u(\cdot, 0)]$.

REMARK 5.3.4. Corollary 5.3.3 generalizes Theorem 5.1.3.

EXERCISE 5.3.5. By imitating the proof of Theorem 5.3.2 or just referring to this theorem show that if $u \in C^{2+\delta}(\mathbb{R}^d_+)$ and $\Delta u \in C^{k+\delta}(\mathbb{R}^d_+)$ and $u(\cdot,0) \in C^{k+2+\delta}(\mathbb{R}^{d-1})$, then $u \in C^{k+2+\delta}(\bar{\mathbb{R}}^d_+)$ and estimate (5.3.5) holds.

EXERCISE 5.3.6. By using Exercise 5.3.5 show that if the function g from Lemma 5.2.1 belongs to $C^{k+2+\delta}(\mathbb{R}^{d-1})$, then $u = \Pi\, g \in C^{k+2+\delta}(\bar{\mathbb{R}}^d_+)$ and

$$[u]_{k+2+\delta;\mathbb{R}^d_+} \leq N[g]_{k+2+\delta;\mathbb{R}^{d-1}}, \quad |u|_{0;\mathbb{R}^d_+} \leq |g|_{0;\mathbb{R}^{d-1}}. \tag{5.3.6}$$

EXERCISE 5.3.7. The number $k+2$ in (5.3.6) varies in the set $\{2,3,...\}$. Show that the result of Exercise 5.3.6 remains true as well when we take 0 or 1 instead of $k+2$.

EXERCISE 5.3.8. By Exercise 5.3.7 and by the interpolation inequalities the simplest of pseudo-differential operators, the Cauchy operator

$$\mathcal{K} : g \to (\Pi\, g)_{x^d}|_{x^d=0}$$

is well defined and bounded as an operator from $C^{k+\delta}(\mathbb{R}^{d-1})$ into $C^{k-1+\delta}(\mathbb{R}^{d-1})$ for any $k \geq 1$. Moreover

$$[\mathcal{K}g]_{k-1+\delta;\mathbb{R}^{d-1}} \leq N[g]_{k+\delta;\mathbb{R}^{d-1}}.$$

Prove that $\mathcal{K}^2 = -\Delta_{d-1}$ on $C^{k+\delta}(\mathbb{R}^{d-1})$ for $k \geq 2$, so that one defines $\sqrt{-\Delta_{d-1}} = -\mathcal{K}$.

EXERCISE 5.3.9. For the operator $\mathcal{K}$ from Exercise 5.3.8, prove that there is a constant $N = N(d,k)$ such that for any $k \geq 0$ and $g \in C_0^\infty(\mathbb{R}^{d-1})$

$$[g]_{k+1+\delta;\mathbb{R}^{d-1}} \leq N[\mathcal{K}g]_{k+\delta;\mathbb{R}^{d-1}}.$$

5.4. Solvability of elliptic equations with variable coefficients in half spaces

Let us go back to our general second–order elliptic operators L. First we repeat some constructions from Sec. 4.1. Recall that $k \geq 1$ is an integer and that $\delta \in (0,1)$.

THEOREM 5.4.1. *Assume that $|a^{ij}, b^i, c|_{k+\delta;\mathbb{R}^d_+} \leq K$, where K is a constant. Then for any $z_0 > 0$ there exists a constant N depending only on $\kappa, \delta, k, K, d, z_0$, such that for any $u \in C_0^{k+2+\delta}(\mathbb{R}^d_+)$ and any $z \geq z_0$ we have*

$$z|u|_{0;\mathbb{R}^d_+} \leq |Lu - zu|_{0;\mathbb{R}^d_+},$$

$$[u]_{k+2+\delta;\mathbb{R}^d_+} \leq N\Big([Lu - zu]_{k+\delta;\mathbb{R}^d_+} + z^{(k+\delta)/2}|Lu - zu|_{0;\mathbb{R}^d_+}\Big). \tag{5.4.1}$$

Proof. Observe that the first estimate in (5.4.1) we have due to Theorem 2.9.2. Next, assume that we have the result for operators with constant coefficients. We claim then that in the general case there are constants $N, z_1 \geq 1$ such that for any $u \in C_0^{2+\delta}(\mathbb{R}_+^d)$ and any $z \geq z_1$ we have (5.4.1). To prove this claim it suffices to repeat word for word our arguments from Sec. 4.1 with only one change: we should take $\mathbb{R}_+^d$ instead of $\mathbb{R}^d$. Once we have estimates (5.4.1) for all $z \geq z_1$, we can go down to smaller $z \geq z_0$ by using Theorem 2.9.2. Indeed,

$$[Lu - z_1 u]_{k+\delta;\mathbb{R}_+^d} \leq [Lu - zu]_{k+\delta;\mathbb{R}_+^d} + |z - z_1|[u]_{k+\delta;\mathbb{R}_+^d}$$

$$\leq [Lu - zu]_{k+\delta;\mathbb{R}_+^d} + \varepsilon[u]_{k+2+\delta;\mathbb{R}_+^d} + N(\varepsilon)|u|_{0;\mathbb{R}_+^d},$$

and moreover $z \geq z_0$, whence $|u|_{0;\mathbb{R}_+^d} \leq z^{-1}|Lu - zu|_{0;\mathbb{R}_+^d} \leq Nz^{(k+\delta)/2}|Lu - zu|_{0;\mathbb{R}_+^d}$. Also

$$z_1^{(k+\delta)/2}|Lu - z_1 u|_{0;\mathbb{R}_+^d} = N|Lu - z_1 u|_{0;\mathbb{R}_+^d} \leq N|Lu - zu|_{0;\mathbb{R}_+^d}$$

$$+ N|u|_{0;\mathbb{R}_+^d} \leq N(1 + z^{-1})|Lu - zu|_{0;\mathbb{R}_+^d} \leq Nz^{(k+\delta)/2}|Lu - zu|_{0;\mathbb{R}_+^d},$$

so that

$$[u]_{k+2+\delta;\mathbb{R}_+^d} \leq N[Lu - zu]_{k+\delta;\mathbb{R}_+^d}$$

$$+ N\varepsilon[u]_{k+2+\delta;\mathbb{R}_+^d} + N(\varepsilon)z^{(k+\delta)/2}|Lu - zu|_{0;\mathbb{R}_+^d},$$

$$[u]_{k+2+\delta;\mathbb{R}_+^d} \leq N\Big([Lu - zu]_{k+\delta;\mathbb{R}_+^d} + z^{(k+\delta)/2}|Lu - zu|_{0;\mathbb{R}_+^d}\Big).$$

Therefore, we can assume that a, b, c are constant. Furthermore, as always dilations show that we can assume as well that $-c + z = 1$. Next after a nondegenerate change of independent variables (see Remark 2.8.1) we can make $L - z$ look like $\Delta + \tilde{b}^j D_j - 1$. If needed, owing to Exercise 2.1.5, we can make an additional orthogonal transformation so that after all transformations the half–space $\mathbb{R}_+^d$ will take its original place in $\mathbb{R}^d$. In this case by Theorem 5.3.2 for any $u \in C_0^{k+2+\delta}(\mathbb{R}_+^d)$ we have

$$|u|_{k+2+\delta;\mathbb{R}_+^d} \leq N|\Delta u - u|_{k+\delta;\mathbb{R}_+^d} \leq N|Lu|_{k+\delta;\mathbb{R}_+^d} + N|u|_{k+1+\delta;\mathbb{R}_+^d}$$

$$\leq N|Lu|_{k+\delta;\mathbb{R}_+^d} + (1/2)|u|_{k+2+\delta;\mathbb{R}_+^d} + N|u|_{0;\mathbb{R}_+^d}.$$

By Theorem 2.9.2 we can replace the last term $N|u|_{0;\mathbb{R}_+^d}$ with $N|Lu|_{k+\delta;\mathbb{R}_+^d}$, and then we get

$$|u|_{k+2+\delta;\mathbb{R}_+^d} \leq N|Lu|_{k+\delta;\mathbb{R}_+^d}.$$

Finally, we leave to the reader to show that linear nondegenerate transformations can only change the constant N in the last estimate. The theorem is proved.

This theorem along with the method of continuity and Theorem 5.3.2 immediately leads us to the following result.

THEOREM 5.4.2. *Let $k \geq 0$ be an integer and assume that $a, b, c \in C^{k+\delta}(\mathbb{R}^d_+)$ and $-c \geq z_0$ with a constant $z_0 > 0$. Then for any $f \in C^{k+\delta}(\mathbb{R}^d_+)$ and $g \in C^{k+2+\delta}(\mathbb{R}^{d-1})$ there exists a unique solution $u \in C^{k+2+\delta}(\bar{\mathbb{R}}^d_+)$ of the equation $Lu(x) = f(x)$, $x \in \mathbb{R}^d_+$, satisfying the boundary condition $u(x', 0) \equiv g(x')$.*

REMARK 5.4.3. If the coefficients of L and f are smoother than C^δ in $B_R \cap \mathbb{R}^d_+$, then the solution u is smoother than $C^{2+\delta}$ in the same domain. This result is contained in Exercise 7.1.5.

REMARK 5.4.4 (cf. Remarks 4.3.5 or 4.4.4). Theorems 5.4.1 and 5.4.2 mean that for any operator L satisfying the conditions of Theorem 5.4.1 and $z > 0$ the operator

$$z - L : C_0^{k+2+\delta}(\mathbb{R}^d_+) \to C^{k+\delta}(\mathbb{R}^d_+)$$

is one-to-one and onto. Moreover, if we denote by $\mathcal{R}_z$ its inverse operator, then for any $f \in C^{k+\delta}(\mathbb{R}^d_+)$ and any $z_0 > 0$ there is a constant N depending only on $\kappa, \delta, k, K, d, z_0$ such that for $z \geq z_0$ and $f \in C^{k+\delta}(\mathbb{R}^d_+)$

$$[\mathcal{R}_z f]_{k+2+\delta;\mathbb{R}^d_+} \leq N z^{(k+\delta)/2} |f|_{k+\delta;\mathbb{R}^d_+}, \quad z|\mathcal{R}_z f|_{0;\mathbb{R}^d_+} \leq |f|_{0;\mathbb{R}^d_+},$$

where the last estimate follows actually from Theorem 2.9.2. This theorem also implies that the operators $\mathcal{R}_z$ are independent of k and δ. Interpolation between these estimates also shows that for $r \leq k + 2 + \delta$

$$[\mathcal{R}_z f]_{r;\mathbb{R}^d_+} \leq z^{r/2-1} N |f|_{k+\delta;\mathbb{R}^d_+}, \quad |\mathcal{R}_z f|_{r;\mathbb{R}^d_+} \leq z^{r/2-1} N |f|_{k+\delta;\mathbb{R}^d_+}.$$

In particular, we have

$$|\mathcal{R}_z f|_{2;\mathbb{R}^d_+} \leq N |f|_{\delta;\mathbb{R}^d_+}, \quad |\mathcal{R}_z f|_{1;\mathbb{R}^d_+} \leq |z|^{-1/2} N |f|_{\delta;\mathbb{R}^d_+},$$

$$|\mathcal{R}_z f|_{1+\delta;\mathbb{R}^d_+} \leq |z|^{(\delta-1)/2} N |f|_{\delta;\mathbb{R}^d_+}, \quad |\mathcal{R}_z f|_{\delta;\mathbb{R}^d_+} \leq |z|^{-1+\delta/2} N |f|_{\delta;\mathbb{R}^d_+}.$$

EXERCISE 5.4.5. Under the assumptions of Theorem 5.4.1 prove that there exists a constant $N = N(d, \kappa, K, k, \delta)$ such that for any $u \in C_0^{k+2+\delta}(\mathbb{R}^d_+)$ we have

$$|u|_{k+2+\delta;\mathbb{R}^d_+} \leq N\Big([a^{ij} u_{x^i x^j}]_{k+\delta;\mathbb{R}^d_+} + |u|_{0;\mathbb{R}^d_+}\Big).$$

5.5. Remarks on the Neumann and other boundary–value problems in half spaces

We have applied Theorem 5.3.2 to investigation of the Dirichlet problem only. However, it can also give us all the necessary starting information on the Neumann problem too. To understand this let us take the operator Π_1 from Lemma 5.2.3 and observe that by Theorem 5.3.2

$$|\Pi_1 g|_{k+2+\delta;\mathbb{R}^d_+} \leq N |g|_{k+2+\delta;\mathbb{R}^{d-1}}.$$

In particular (remember that Δ_{d-1} is Laplace's operator with respect to $x^1, ..., x^{d-1}$),

$$|\Pi_1(\Delta_{d-1} - 1)g|_{k+\delta;\mathbb{R}^d_+} = |(\Delta_{d-1} - 1)\Pi_1 g|_{k+\delta;\mathbb{R}^d_+} \leq N |(\Delta_{d-1} - 1)g|_{k+\delta;\mathbb{R}^{d-1}}.$$

Since $(\Delta_{d-1}-1)C^{k+2+\delta}(\mathbb{R}^{d-1})=C^{k+\delta}(\mathbb{R}^{d-1})$, we conclude

$$|\Pi_1 g|_{k+\delta;\mathbb{R}^d_+}\leq N|g|_{k+\delta;\mathbb{R}^{d-1}}$$

for any integer $k\geq 0$ with N independent of $g\in C^{k+\delta}(\mathbb{R}^{d-1})$. This inequality shows that for one of the simplest *pseudo-differential* operators $\mathcal{K}_1 g:=(\Pi_1 g)_{x^d}|_{x^d=0}$ we have

$$|\mathcal{K}_1 g|_{k+\delta;\mathbb{R}^{d-1}}\leq N|g|_{k+1+\delta;\mathbb{R}^{d-1}}. \tag{5.5.1}$$

Furthermore, let us find $\mathcal{K}_1^2 g$ for $g\in C^{2+\delta}(\mathbb{R}^{d-1})$. Denote $h=\mathcal{K}_1 g$; then $\Pi_1 h$ is a unique solution of class $C^{2+\delta}_{loc}(\mathbb{R}^d_+)\cap C^{1+\delta}(\mathbb{R}^d_+)$ of the equation $\Delta u-u=0$ with the boundary value h. Since $h=(\Pi_1 g)_{x^d}|_{x^d=0}$ and $(\Pi_1 g)_{x^d}$ is also a bounded solution of the same boundary–value problem, we have $\Pi_1 h=(\Pi_1 g)_{x^d}$. Therefore,

$$\mathcal{K}_1^2 g=\mathcal{K}_1 h=(\Pi_1 h)_{x^d}|_{x^d=0}=(\Pi_1 g)_{x^d x^d}|_{x^d=0}.$$

But $\Delta\Pi_1 g-\Pi_1 g=0$ in $\mathbb{R}^d_+$ and in $\bar{\mathbb{R}}^d_+$. In particular, this equality holds for $x^d=0$. Upon observing that $\Pi_1 g=g$ for $x^d=0$, we conclude

$$\mathcal{K}_1^2 g+\Delta_{d-1}g-g=(\Delta\Pi_1 g-\Pi_1 g)|_{x^d=0}=0,\quad \mathcal{K}_1^2=1-\Delta_{d-1}.$$

(One can find further discussion of the last formula in Exercise 9.1.7.) This and the bijectivity of $1-\Delta_{d-1}$ imply that

$$\mathcal{K}_1\Big(C^{k+1+\delta}(\mathbb{R}^{d-1})\Big)=C^{k+\delta}(\mathbb{R}^{d-1}).$$

This means that for any $h\in C^{k+\delta}(\mathbb{R}^{d-1})$ we can find $g\in C^{k+1+\delta}(\mathbb{R}^{d-1})$ such that $\mathcal{K}_1 g=h$.

It follows, in particular, that for any $g\in C^{k+1+\delta}(\mathbb{R}^{d-1})$ we have

$$|g|_{k+1+\delta;\mathbb{R}^{d-1}}\leq N|\mathcal{K}_1 g|_{k+\delta;\mathbb{R}^{d-1}}. \tag{5.5.2}$$

Indeed, find $h\in C^{k+2+\delta}(\mathbb{R}^{d-1})$ such that $\mathcal{K}_1 h=g$ and use (5.5.1). Then

$$|g|_{k+1+\delta;\mathbb{R}^{d-1}}=|\mathcal{K}_1 h|_{k+1+\delta;\mathbb{R}^{d-1}}\leq N|h|_{k+2+\delta;\mathbb{R}^{d-1}}$$

$$\leq N|(1-\Delta_{d-1})h|_{k+\delta;\mathbb{R}^{d-1}}=N|\mathcal{K}_1 g|_{k+\delta;\mathbb{R}^{d-1}}.$$

Upon solving the equation $\mathcal{K}_1 g=h$ and denoting $u=\Pi_1 g$ we also conclude that

if $r\geq 0$ and $h\in C^{1+r+\delta}(\mathbb{R}^{d-1})$, there exists a unique solution $u\in C^{2+r+\delta}(\mathbb{R}^d_+)$ of the equation $\Delta u-u=0$ such that ($u=g$ and) $u_{x^d}=h$ on $\mathbb{R}^{d-1}$. In addition, by (5.3.4) and (5.5.2)

$$|u|_{r+2+\delta;\mathbb{R}^d_+}\leq N|g|_{r+2+\delta;\mathbb{R}^{d-1}}\leq N|\mathcal{K}_1 g|_{r+1+\delta;\mathbb{R}^{d-1}}=N|h|_{1+r+\delta;\mathbb{R}^{d-1}},$$

$$|u|_{r+2+\delta;\mathbb{R}^d_+}\leq N|u_{x^d}|_{r+1+\delta;\mathbb{R}^{d-1}}.$$

In this way we get a Neumann–problem counterpart of Theorem 5.3.2. One might also be interested in the oblique derivative boundary–value problem, in which one is required to find a function u satisfying

$$\Delta u - u = 0 \quad \text{in} \quad \mathbb{R}^d_+, \quad \frac{\partial u}{\partial b} = h \quad \text{on} \quad \mathbb{R}^{d-1}, \tag{5.5.3}$$

where b is a given vector with $b^d = 1$. In this situation one has to consider the equation $b^1 g_{x^1} + ... + b^{d-1} g_{x^{d-1}} + \mathcal{K}_1 g = h$ in $\mathbb{R}^{d-1}$ and then set $u = \Pi_1 g$. Formally, in the Fourier images this equation says that $[ib' \cdot \xi - \sqrt{|\xi|^2 + 1}]\tilde{g}(\xi) = \tilde{h}(\xi)$, $\xi \in \mathbb{R}^{d-1}$, and one has a justification to repeat all the above theory starting from Sec. 1.2, this time taking as $p(\xi)$ not a polynomial but $ib' \cdot \xi - \sqrt{|\xi|^2 + 1}$. This way of investigating (5.5.3) can be generalized for the case in which instead of the second condition in (5.5.3) we have

$$\sum_{i,j=1}^{d-1} \alpha^{ij} \frac{\partial^2 u}{\partial x^i \partial x^j} + \frac{\partial u}{\partial b} = h, \tag{5.5.4}$$

where (α^{ij}) is a symmetric strictly positive $(d-1) \times (d-1)$ matrix independent of x. Indeed, in this case in the Fourier images we get $[-\sum_{i,j=1}^{d-1} \alpha^{ij} \xi^i \xi^j + ib' \cdot \xi - \sqrt{|\xi|^2 + 1}]\tilde{g}(\xi) = \tilde{h}(\xi)$, $\xi \in \mathbb{R}^{d-1}$. In this situation one can solve the boundary–value problem for less regular functions: $h \in C^\delta(\mathbb{R}^{d-1})$. Actually if one considers

$$\sum_{i,j=1}^{d-1} \alpha^{ij} \frac{\partial^2 u}{\partial x^i \partial x^j} + \frac{\partial u}{\partial b} - u = h$$

instead of (5.5.4), then appealing to the Fourier transform is not necessary, since one can solve the equation

$$\sum_{i,j=1}^{d-1} \alpha^{ij} \frac{\partial^2 g}{\partial x^i \partial x^j} + \sum_{i=1}^{d-1} b^i g_{x^i} + \mathcal{K}_1 g - g = h$$

on $\mathbb{R}^{d-1}$ just as above by treating $\mathcal{K}_1 g$ in the same way as first–order terms (cf. proof of Theorem 5.4.1) and using the method of continuity.

Further generalization of the boundary condition (5.5.4) consists in considering

$$L'u + u_{x^d} = h,$$

where L' is an mth order elliptic operator with constant coefficients in the space $\mathbb{R}^{d-1}$. In this case one finds a solution u as $\Pi_1 g$, where g satisfies $L'g + \mathcal{K}_1 g = h$. An important advantage of boundary–value conditions like (5.5.3) and (5.5.4) is that there is a kind of maximum principle for their solutions (if $h \geq 0$, then the bounded solution of $\Delta u - u = 0$ with the boundary condition either (5.5.3) or (5.5.4) is nonpositive and, moreover, $|g|_{0;\mathbb{R}^{d-1}} \leq |h|_{0;\mathbb{R}^{d-1}}$).

It is also worth noting that there are *explicit* representations of solutions to (5.5.3) which allow one to get the solvability and a priori estimates (see, for instance, [**6**]).

One more way of treating boundary–value problems like (5.5.3) is to differentiate the equation. Let us consider (5.5.3) and assume that $h \in C^{1+\delta}(\mathbb{R}^{d-1})$ and a function $u \in C^{2+\delta}(\mathbb{R}^d_+)$ satisfies (5.5.3). Then u is infinitely differentiable in $\mathbb{R}^d_+$, and the function $v := \partial u/\partial b$ defined in $\bar{\mathbb{R}}^d_+$ satisfies the equation $\Delta v - v = 0$ in $\mathbb{R}^d$ and $v = h$ on $\mathbb{R}^{d-1}$. By the above $v = \Pi_1 h$ and

$$|v|_{1+\delta;\mathbb{R}^d_+} \le N|h|_{1+\delta;\mathbb{R}^{d-1}}. \tag{5.5.5}$$

Simple computations show that (remember $b^d = 1$)

$$u_{x^dx^d} = \sum_{i,j=1}^{d-1} b^ib^ju_{x^ix^j} + 2\sum_{i=1}^{d} b^iu_{x^ix^d} - b^ib^ju_{x^ix^j} = \sum_{i,j=1}^{d-1} b^ib^ju_{x^ix^j} + 2v_{x^d} - b^iv_{x^i}.$$

In particular on $\mathbb{R}^{d-1} = \partial\mathbb{R}^d_+$ the function u satisfies the following elliptic equation

$$\Delta_{d-1}u + \sum_{i,j=1}^{d-1} b^ib^ju_{x^ix^j} - u = b^iv_{x^i} - 2v_{x^d}.$$

It follows that the $C^{2+\delta}(\mathbb{R}^{d-1})$-norm of $u(\cdot,0)$ is controlled by $C^{1+\delta}(\mathbb{R}^d_+)$–norm of v. Finally, taking into account (5.5.5) and that $u = \Pi_1 u(\cdot,0)$, we get the following a priori estimate:

$$|u|_{2+\delta;\mathbb{R}^d_+} \le N|h|_{1+\delta;\mathbb{R}^{d-1}}.$$

The only inconvenience of this proof in comparison to, say, Lemma 5.2.3 is that we do not know if the solution u exists at all. But once we have obtained solvability of the Neumann boundary-value problem, we can prove solvability of (5.5.3) applying the method of continuity when instead of b in (5.5.3) one takes $tb + (1-t)e_d$.

5.6. Hints to exercises

5.1.1. Remember that u is infinitely differentiable in Ω so that one can differentiate the equation $\Delta u = 0$.
5.2.4. Show that $u(0,y)$ behaves like $y\log(1/y)$ as $y \to 0$.
5.3.7. First assume that $g \in C^\infty_b(\mathbb{R}^{d-1})$. Then observe that

$$[\Delta_{d-1}u]_{k+\delta;\mathbb{R}^d_+} \le N[u]_{k+2+\delta;\mathbb{R}^d_+}, \quad [g]_{k+2+\delta;\mathbb{R}^{d-1}} \le N[\Delta_{d-1}g]_{k+\delta;\mathbb{R}^{d-1}}.$$

After this use Exercise 1.3.4 and the fact that Δ_{d-1} and Π commute in order to get the result for $g \in C^\infty_0(\mathbb{R}^d)$. Finally, use Exercises 3.2.3 and 3.2.4.
5.3.8. Observe that $(\Pi\, g)_{x^d}$ is a bounded harmonic function with boundary values $(\Pi\, g)_{x^d}|_{x^d=0}$.
5.3.9. For $g = \mathcal{K}h$ get the estimate from the previous exercise. Then use this exercise again along with Exercise 1.3.4 to approximate by $\mathcal{K}h$, even by $\mathcal{K}^2h$, any function $g \in C^\infty_0(\mathbb{R}^{d-1})$.

CHAPTER 6

Second–Order Elliptic Equations in Smooth Domains

Throughout this chapter k is an integer, $k \geq 0$, $\delta \in (0,1)$. We also take a constant $K > 0$ and a second–order elliptic operator L with the coefficients a, b, c $(c \leq 0)$ and the ellipticity constant $\kappa > 0$ and assume that $|a, b, c|_{k+\delta} \leq K$. In particular, we assume that $a, b, c \in C^{k+\delta}(\mathbb{R}^d)$.

6.1. The maximum principle. Domains of class C^r

We start with the maximum principle for bounded domains, when the restriction $c < 0$ is not required, and with the general notion of smooth domain. We need the following lemma about existence of so–called *global barriers*.

LEMMA 6.1.1. *For any $R > 0$ there exists a function $v_0 \in C^\infty(\bar{B}_R)$ such that $Lv_0 \leq -1$ in B_R. Moreover, $0 < v_0 \leq N_0 = N_0(\kappa, K, R, d)$ in B_R and $v_0 = 0$ on ∂B_R.*

Proof. Take $v = \cosh(\lambda|x|)$, $\lambda > 0$. By considering Taylor's expansion of cosh one sees that v is infinitely differentiable. Also for $x \neq 0$

$$D_i v = \lambda \sinh(\lambda|x|)\xi^i \quad \text{where} \quad \xi = x|x|^{-1},$$

$$D_i D_j v = \lambda^2 \cosh(\lambda|x|)\xi^i\xi^j + \lambda|x|^{-1}\sinh(\lambda|x|)(\delta^{ij} - \xi^i\xi^j).$$

Bearing in mind that $|\xi| = 1$ and $\sinh t < \cosh t$ and $\operatorname{tr} a \geq a^{ij}\xi^i\xi^j$, for $x \neq 0$ we obtain

$$(L - c)v = (a^{ij}D_iD_j + b^iD_i)\cosh(\lambda|x|) \geq$$

$$\lambda^2\cosh(\lambda|x|)a^{ij}\xi^i\xi^j + \lambda\sinh(\lambda|x|)b^i\xi^i \geq \cosh(\lambda|x|)(\lambda^2\kappa - \lambda N) \geq 1$$

under an appropriate choice of $\lambda = \lambda(\kappa, K, d)$. The inequality $(L - c)v \geq 1$ is also true for $x = 0$ since v is a C^∞–function.

Now take $v_0(x) = \cosh(\lambda R) - v(x)$. Then, of course, $0 < v_0 \leq N_0$ in B_R and $v = 0$ on ∂B_R. Moreover, since $c \leq 0$,

$$Lv_0 \leq (L - c)v_0 = -(L - c)v \leq -1$$

in B_R, and the lemma is proved.

COROLLARY 6.1.2 (the maximum principle). *Let Ω be a bounded domain. If $u \in C^2_{loc}(\Omega) \cap C(\bar{\Omega})$ and $Lu \leq 0$ in Ω and $u \geq 0$ on $\partial\Omega$, then $u \geq 0$ in Ω.*

To prove this without loss of generality we may assume that $0 \in \Omega$. Then define $R = 2\operatorname{diam}\Omega$, and take the function v_0 from Lemma 6.1.1. Also define the operator $L'w = L(v_0 w)$ and call a', b', c' its coefficients. It is easy to see that L' is an elliptic operator and $c' \le -1$. Also the function $\bar{u} := u/v_0$ is continuous in $\bar{\Omega}$ and $L'\bar{u} = Lu \le 0$ in Ω and $\bar{u} \ge 0$ on $\partial\Omega$. By Theorem 2.9.1 we have $\bar{u} \ge 0$ and $u \ge 0$ in Ω.

EXERCISE 6.1.3. Let Ω be a bounded domain and $u \in C^2_{loc}(\Omega) \cap C(\bar{\Omega})$. Define $f = Lu$ and assume that $|f| \le -c$ and $u = 0$ on $\partial\Omega$. By comparing u and $v \equiv 1$ prove that $|u| \le 1$.

EXERCISE* 6.1.4. By considering $L(\rho^2 - |x|^2)$ prove that there exist constants N, ρ_1 depending only on κ, K, d such that if $\rho \in (0, \rho_1]$ and $u \in C^2(B_\rho) \cap C(\bar{B}_\rho)$ and $Lu \ge -1$ in B_ρ and $u \le 0$ on ∂B_ρ, then $u(x) \le N(\rho^2 - |x|^2)$ in B_ρ. Also prove that if $\rho \in (0,1]$ and $u \in C^2(B_\rho) \cap C(\bar{B}_\rho)$ and $Lu \ge -1$ in B_ρ and $u \le 0$ on ∂B_ρ, then $u(x) \le N\rho^2$ in B_ρ, where N depends only on κ, K, d.

EXERCISE* 6.1.5. Let $R \in (0,1]$, $r \in (0,1)$, $r < \rho < 1$, $\Omega = B_R \setminus \bar{B}_{rR}$. Prove that there is a constant $\gamma = \gamma(d, K, \kappa, r, \rho) < 1$ (independent of R) such that if $u \in C^2_{loc}(\Omega) \cap C(\bar{\Omega})$ and $Lu \ge 0$ in Ω and $u \le 1$ on ∂B_R and $u \le 0$ on ∂B_{rR}, then $u(x) \le \gamma$ for $rR \le |x| \le \rho R$.

In what follows in this chapter we deal only with smooth domains Ω.

DEFINITION 6.1.6. Let $r > 0$ and Ω be a bounded domain in $\mathbb{R}^d$. We write $\Omega \in C^r$ (or $\partial\Omega \in C^r$) and say that the domain Ω is of class C^r if there are numbers $K_0, \rho_0 > 0$ such that for any point $x_0 \in \partial\Omega$ there exists a one-to-one mapping ψ of $B_{\rho_0}(x_0)$ onto a domain $D \subset \mathbb{R}^d$ such that

(i) $D_+ := \psi(B_{\rho_0}(x_0) \cap \Omega) \subset \mathbb{R}^d_+$ and $\psi(x_0) = 0$,

(ii) $\psi(B_{\rho_0}(x_0) \cap \partial\Omega) = D \cap \{y \in \mathbb{R}^d : y^d = 0\}$,

(iii) $[\psi]_{s;B_{\rho_0}(x_0)} + [\psi^{-1}]_{s;D} \le K_0$ for any $s \in [0,r]$, and $|\psi^{-1}(y_1) - \psi^{-1}(y_2)| \le K_0|y_1 - y_2|$ for any $y_i \in D$.

We say that the diffeomorphism ψ *straightens the boundary* near x_0.

EXERCISE 6.1.7. Show that in $\mathbb{R}^3 = \{(x,y,z)\}$ the domain $x^2 + y^2 + z^2 < 1$ is of class C^r for any r.

In the future we always assume that $r = k+2+\delta$. We also fix a domain $\Omega \in C^r$.

Let $v \in C^r(B_{\rho_0}(x_0) \cap \Omega)$. In the domain D_+ define the function $w(y) = v(x)$, where $x = \psi^{-1}(y)$. Obviously, $v(x) = w(y)$ in $B_{\rho_0}(x_0) \cap \Omega$ for $y = \psi(x)$, and

$$v_{x^i}(x) = w_{y^k}(y)\psi^k_{x^i}(x), \quad v_{x^i x^j}(x) = w_{y^k}(y)\psi^k_{x^i x^j}(x) + w_{y^k y^l}(y)\psi^k_{x^i}(x)\psi^l_{x^j}(x).$$

Like formulas immediately lead to the following lemma.

LEMMA 6.1.8. $v \in C^r(B_{\rho_0}(x_0) \cap \Omega) \iff w \in C^r(D_+)$. *In addition,*

$$|v|_{r;B_{\rho_0}(x_0)\cap\Omega} \le N|w|_{r;D_+}, \quad |w|_{r;D_+} \le N|v|_{r;B_{\rho_0}(x_0)\cap\Omega},$$

where $N = N(d, r, K_0)$.

EXERCISE* 6.1.9. Prove Lemma 6.1.8.

6.2. Equations near the boundary

A rough idea of how to solve the equation $Lu = f$ in a domain Ω is to solve it locally in neighborhoods of any boundary point by straightening the relevant piece of the boundary. We need to understand how the mapping ψ affects the equation.

Let $v \in C^r(B_{\rho_0}(x_0) \cap \Omega)$. Assume that in $B_{\rho_0}(x_0) \cap \Omega$

$$Lv(x) = a^{ij}(x)v_{x^i x^j}(x) + b^i(x)v_{x^i}(x) + c(x)v(x) = f(x). \tag{6.2.1}$$

From the computations in the previous section we see that $(y = \psi(x), w(y) = v(x))$

$$\tilde{L}w(y) := \tilde{a}^{kl}(y)w_{y^k y^l}(y) + \tilde{b}^k(y)w_{y^k}(y) + \tilde{c}(y)w(y) = \tilde{f}(y), \tag{6.2.2}$$

where

$$\tilde{a}^{kl}(y) = a^{ij}(x)\psi^k_{x^i}(x)\psi^l_{x^j}(x), \quad \tilde{b}^k(y) = a^{ij}(x)\psi^k_{x^i x^j}(x) + b^i(x)\psi^k_{x^i}(x),$$

$$\tilde{c}(y) = c(x), \quad \tilde{f}(y) = f(x).$$

LEMMA 6.2.1. *For any ψ and D from Definition 6.1.6 we have $|\tilde{a}, \tilde{b}, \tilde{c}|_{k+\delta;D} \le N(d, K_0, K, k, \delta)$. For a twice continuously differentiable in $B_{\rho_0}(x_0) \cap \Omega$ function v equality (6.2.1) holds in $B_{\rho_0}(x_0) \cap \Omega$ if and only if (6.2.2) holds in D_+ for w. Furthermore, $\tilde{L}$ is an elliptic operator in D and*

$$\tilde{a}^{ij}(y)\xi^i\xi^j \ge \tilde{\kappa}|\xi|^2, \quad \forall \xi \in \mathbb{R}^d, y \in D,$$

where $\tilde{\kappa} = \tilde{\kappa}(d, K_0, \kappa) > 0$.

Proof. The first assertion follows from Lemma 6.1.8. The second one has been checked out in one direction. To prove the remaining part it suffices to *define* $v(x) = w(y), g = Lv$, notice that by what has been already proved, $Lw = \tilde{g}$, where $\tilde{g}(y) = g(x)$, and finally remember that $Lw = \tilde{f}$. To prove the last assertion of the lemma we notice that

$$\tilde{a}^{kl}(y)\xi^k\xi^l = a^{ij}(x)(\xi \cdot \psi)_{x^i}(x)(\xi \cdot \psi)_{x^j}(x) \ge \kappa|(\xi \cdot \psi)_x(x)|^2,$$

and for $\phi := \psi^{-1}$ we have

$$\psi^k_{x^i}(x)\phi^i_{y^j}(y) = \delta^k_j, \quad (\xi \cdot \psi)_{x^i}(x)\phi^i_{y^j}(y) = \xi^j,$$

so that $|\xi| \le NK_0|(\xi \cdot \psi)_x|$. The lemma is proved.

Now we fix a function $\eta \in C_0^\infty(\mathbb{R}^d)$ such that $\eta(x) = 1$ for $|x| \le \rho_0/2$ and $\eta(x) = 0$ for $|x| \ge \rho_0$ and $0 \le \eta \le 1$, and we define

$$\tilde{\eta}(y) = \eta(x - x_0), \quad \bar{L}(y) = \tilde{\eta}(y)\tilde{L}(y) + (1 - \tilde{\eta}(y))\Delta.$$

The operator $\bar{L}$ is elliptic together with $\tilde{L}$. One of its advantages is that it is defined for all $y \in \mathbb{R}^d_+$. This and Remark 5.4.4 allow us to define the operators $\bar{\mathcal{R}}_z, z > 0$, as the inverse operators to

$$z - \bar{L} : C_0^{k+2+\delta}(\mathbb{R}^d_+) \to C^{k+\delta}(\mathbb{R}^d_+).$$

Also, we define the operators

$$\Psi : w = w(y) \to \Psi w(x) = w(\psi(x)),\ \Psi^{-1} : v = v(x) \to \Psi^{-1} v(y) = v(\psi^{-1}(y)),$$

$$S_z^{x_0} : f = f(x) \to S_z^{x_0} f(x) = \Psi \bar{\mathcal{R}}_z \Psi^{-1}[\eta(\cdot - x_0) f](x),$$

where by $\Psi^{-1}[\eta(\cdot - x_0)f](y)$ we certainly mean the function $\eta(\psi^{-1}(y) - x_0) f(\psi^{-1}(y))$ for $y \in D_+$ and zero outside D_+. Finally, define

$$C_0^{2+\delta}(\Omega) = C^{2+\delta}(\bar{\Omega}) \cap \{u : u = 0 \text{ on } \partial\Omega\}.$$

THEOREM 6.2.2. *(i) If $v \in C_0^{2+\delta}(\Omega)$ and $v = 0$ outside $B_{\rho_0/2}(x_0) \cap \Omega$, then in $B_{\rho_0/2}(x_0) \cap \Omega$ for any $z > 0$ we have*

$$v = S_z^{x_0}(zv - Lv).$$

(ii) There is a constant N depending only on $\kappa, K_0, \rho_0, \delta, k, K, d$ such that for $z \geq 1$ and $f \in C^{k+\delta}(\Omega)$ we have $S_z^{x_0} f \in C^{k+2+\delta}(B_{\rho_0}(x_0) \cap \Omega)$ and

$$|S_z^{x_0} f|_{k+2+\delta; B_{\rho_0}(x_0)\cap\Omega} \leq N z^{(k+\delta)/2} |f|_{k+\delta; B_{\rho_0}(x_0)\cap\Omega},$$

$$|S_z^{x_0} f|_{1+\delta; B_{\rho_0}(x_0)\cap\Omega} \leq N z^{(\delta-1)/2} |f|_{\delta; B_{\rho_0}(x_0)\cap\Omega}. \tag{6.2.3}$$

Proof. (i) Define $f = zv - Lv$ and introduce w as above. By Lemma 6.2.1 we have $\tilde{f} = \Psi^{-1} f = zw - \tilde{L}w$ in D_+. Since $v \neq 0$ only when $\eta(\cdot - x_0) = 1$, also $w \neq 0$ only if $\tilde{\eta} = 1$. It follows that

$$\tilde{f} = zw - \bar{L}w$$

in D_+. Actually this equality holds in $\mathbb{R}^d_+$ if we continue $\tilde{f}, w$ as zero outside D_+. Also $\eta(x - x_0) f(x) = f(x)$, so that $\tilde{f} = \Psi^{-1} f = \Psi^{-1}[\eta(\cdot - x_0) f]$. For $y \in D_+$ and $x = \psi^{-1}(y) \in B_{\rho_0/2}(x_0) \cap \Omega$ this yields

$$w(y) = \bar{\mathcal{R}}_z \tilde{f}(y) = \bar{\mathcal{R}}_z \Psi^{-1}[\eta(\cdot - x_0) f](y),$$

$$v(x) = \Psi w(x) = \Psi \bar{\mathcal{R}}_z \Psi^{-1}[\eta(\cdot - x_0) f](x) = S_z^{x_0} f(x).$$

(ii) The inclusion $S_z^{x_0} f \in C^{k+2+\delta}(B_{\rho_0}(x_0) \cap \Omega)$ is known from Lemma 6.1.8 and from Remark 5.4.4, which also imply that the left–hand side of the second inequality in (6.2.3) is less than

$$N z^{(\delta-1)/2} |\Psi^{-1}[\eta(\cdot - x_0) f]|_{\delta; \mathbb{R}^d_+} = N z^{(\delta-1)/2} |\Psi^{-1}[\eta(\cdot - x_0) f]|_{\delta; D_+}.$$

One more application of Lemma 6.1.8 proves the second inequality in (6.2.3). One proves the first one in like manner. The theorem is proved.

REMARK 6.2.3. Assertion (i) of Theorem 6.2.2 says that if we want to solve the equation $zu - Lu = f$ in $C_0^{2+\delta}(\Omega)$ and we know in advance that the solution vanishes outside $B_{\rho_0/2}(x_0) \cap \Omega$, then the solution in $B_{\rho_0/2}(x_0) \cap \Omega$ is given by $S_z^{x_0} f$.

EXERCISE* 6.2.4. Prove that if $f \in C^{\delta}(B_{\rho_0}(x_0) \cap \Omega)$, then

$$S_z^{x_0} f \in C^{2+\delta}(B_{\rho_0}(x_0) \cap \Omega), \quad S_z^{x_0} f = 0 \quad \text{on} \quad B_{\rho_0}(x_0) \cap \partial\Omega$$

and $(z - L)S_z^{x_0} f = f$ in $B_{\rho_0/2}(x_0) \cap \Omega$.

6.3. Partitions of unity and a priori estimates

Take a bounded domain $\Omega \in C^{k+2+\delta}$. In the previous section we saw that some results concerning elliptic operators near $\partial\Omega$ can be obtained by straightening the boundary. We need a procedure allowing us to get certain results for entire Ω from similar results obtained for half spaces and the whole space.

Take a function $\xi \in C_0^\infty(\mathbb{R}^d)$ such that $\xi = 0$ for $|x| \geq \rho_0/2$, $\xi(x) = 1$ for $|x| \leq \rho_0/4$ and $0 \leq \xi \leq 1$. Next, take points $x_1, x_2, ... \in \partial\Omega$ so that $|x_i - x_j| \geq \rho_0/8$ for $i \neq j$ and the whole $\partial\Omega$ is covered by $B_{\rho_0/8}(x_i)$. The number of the points x_i needed to carry out this construction is finite. We denote by n this number. Observe that n can be estimated through ρ_0, d and the diameter of Ω, which we denote by d_Ω.

Define $\xi^i(x) = \xi(x - x_i)$. To complete the system of these functions, one can find a function $\xi^0 \in C_0^\infty(\mathbb{R}^d)$, $0 \leq \xi^0 \leq 1$, such that $\xi^0(x) = 0$ for $x \in \Omega$ with $\text{dist}\,(x, \partial\Omega) \leq \rho_0/16$ and $\xi^0(x) = 1$ if $x \in \Omega$ and $\text{dist}\,(x, \partial\Omega) \geq \rho_0/8$. One can manage to do this, for instance, by mollifying the indicator of $\Omega \setminus \{x : \text{dist}\,(x, \partial\Omega) \leq 3\rho_0/32\}$.

Notice that $\sum_{i \geq 1} (\xi^i)^2(x)$ is greater that 1 if $x \in \bar{\Omega}$ and $\text{dist}\,(x, \partial\Omega) \leq \rho_0/8$. Therefore, the function

$$\bar{\xi} = \sum_{i \leq n} (\xi^i)^2$$

is greater than 1 in $\bar{\Omega}$. Also, $\bar{\xi}$ and every derivative of it of any order are bounded in $\bar{\Omega}$ by a number depending only on d, n and the order of the derivative. Finally, define $\zeta^i = \xi^i \bar{\xi}^{-1/2}, i \geq 0$, and notice that all ζ^i are infinitely differentiable and in $\bar{\Omega}$

$$\sum_{i \leq n} (\zeta^i)^2 = 1.$$

We have constructed a so-called partition of unity in Ω.

The partition of unity allows us to carry the interpolation inequalities over the case of functions in domains. In the following lemma we assume that $\Omega \in C^{k+2+\delta}$ just for convenience. Actually its statement is true for Lipschitz domains.

LEMMA 6.3.1. *If $r = k + 2 + \delta \geq s \geq 0$, then there exists a constant $N = N(d, k, \delta, K_0, \rho_0, s, d_\Omega)$ such that for any $v \in C^r(\Omega)$ we have*

$$|v|_{s;\Omega} \leq N\Big(|v|_{r;\Omega}^{s/r} |v|_{0;\Omega}^{1-s/r} + |v|_{0;\Omega}\Big).$$

In particular, for any $\varepsilon > 0$

$$|v|_{s;\Omega} \leq N\varepsilon^{r-s} |v|_{r;\Omega} + N(1 + \varepsilon^{-s})|v|_{0;\Omega}.$$

Proof. At first let $s = p + \theta$, where $\theta \in (0,1)$ and p is an integer. Let D^i, Ψ_i be the domain D and operator Ψ corresponding to x_i, and let $w_i = \Psi_i^{-1} v$, $\tilde{w}_i = \Psi_i^{-1}[(\zeta^i)^2 v]$. For $i = 1, ..., n$ we have (with the first inequality following from Exercise 3.1.6)

$$|(\zeta^i)^2 v|_{s;\Omega} \leq N|(\zeta^i)^2 v|_{s;B_{\rho_0}(x_i)\cap\Omega} \leq N|\tilde{w}_i|_{s;D^i_+} \leq N|\tilde{w}_i|_{s;\mathbb{R}^d_+}$$

$$= N\sum_{j=1}^{p}[\tilde{w}_i]_{j;\mathbb{R}^d_+} + N[\tilde{w}_i]_{s;\mathbb{R}^d_+} \leq N\sum_{j=1}^{p}[\tilde{w}_i]^{j/r}_{r;\mathbb{R}^d_+}|\tilde{w}_i|^{1-j/r}_{0;\mathbb{R}^d_+} + [\tilde{w}_i]^{s/r}_{r;\mathbb{R}^d_+}|\tilde{w}_i|^{1-s/r}_{0;\mathbb{R}^d_+}$$

$$\leq N\sum_{j=1}^{p}|w_i|^{j/r}_{r;D^i_+}|w_i|^{1-j/r}_{0;D^i_+} + |w_i|^{s/r}_{r;D^i_+}|w_i|^{1-s/r}_{0;D^i_+}$$

$$\leq N\left(1 + \left[\frac{|w_i|_{r;D^i_+}}{|w_i|_{0;D^i_+}}\right]^{s/r}\right)|w_i|_{0;D^i_+} = N\Big(|w_i|^{s/r}_{r;D^i_+}|w_i|^{1-s/r}_{0;D^i_+} + |w_i|_{0;D^i_+}\Big)$$

$$\leq N\Big(|v|^{s/r}_{r;\Omega}|v|^{1-s/r}_{0;\Omega} + |v|_{0;\Omega}\Big).$$

The same kind of estimate is true for $|(\zeta^0)^2 v|_{s;\Omega}$. Also the case in which s is an integer is considered in like manner. It remains only to observe that $v = \sum_{i\leq n}(\zeta^i)^2 v$. The lemma is proved.

Our next step consists of proving an a priori estimate.

THEOREM 6.3.2. *There is a constant N depending only on $d_\Omega, k, \delta, d, \kappa, K, K_0, \rho_0$ such that for any $u, g \in C^{k+2+\delta}(\Omega)$, satisfying $u = g$ on $\partial\Omega$, we have*

$$|u|_{k+2+\delta;\Omega} \leq N\Big(|Lu|_{k+\delta;\Omega} + |g|_{k+2+\delta;\Omega}\Big). \tag{6.3.1}$$

Proof. Obviously (and by Lemma 6.3.1)

$$|u|_{k+2+\delta;\Omega} \leq |u-g|_{k+2+\delta;\Omega} + |g|_{k+2+\delta;\Omega}, \quad |L(u-g)|_{k+\delta;\Omega} \leq |Lu|_{k+\delta;\Omega} + N|g|_{k+2+\delta;\Omega}.$$

It follows that we need consider only $u - g$ or that we may assume that $u \in C_0^{k+2+\delta}(\Omega)$. In that case if $i \geq 1$ and we denote $S^i_z = S^{x_i}_z$, then by Theorem 6.2.2

$$|(\zeta^i)^2 u|_{k+2+\delta;\Omega} \leq N|\zeta^i u|_{k+2+\delta;B_{\rho_0/2}(x_i)\cap\Omega} = N|S^i_1(\zeta^i u - L(\zeta^i u))|_{k+2+\delta;B_{\rho_0/2}(x_i)\cap\Omega}$$

$$\leq N|\zeta^i u - L(\zeta^i u)|_{k+\delta;B_{\rho_0}(x_i)\cap\Omega} \leq N\Big(|Lu|_{k+\delta;\Omega} + |u|_{k+2;\Omega}\Big).$$

By Theorem 4.3.2 the same estimate is true for $i = 0$. We sum up these estimates and apply Lemma 6.3.1 to arrive at

$$|u|_{k+2+\delta;\Omega} \leq N\Big(|Lu|_{k+\delta;\Omega} + |u|_{0;\Omega}\Big).$$

In order to get (6.3.1) it suffices now to estimate $|u|_{0;\Omega}$ through $|Lu|_{k+\delta;\Omega}$. Without loss of generality we assume that $0 \in \Omega$. Then define $R = 2d_\Omega$, and take the function v_0 from Lemma 6.1.1. For the function $w = u - v_0|Lu|_{0;\Omega}$ we obviously have $Lw \geq 0$ in Ω and $w \leq 0$ on $\partial\Omega$. By Corollary 6.1.2 we get $u \leq v_0|Lu|_{0;\Omega}$ in Ω. Similarly, $-u \leq v_0|Lu|_{0;\Omega}$, so that $|u|_{0;\Omega} \leq |v_0|_{0;\Omega}|Lu|_{0;\Omega}$, and the theorem is proved.

REMARK 6.3.3. If we knew that the equation $\Delta u - u = f$ is solvable in $C_0^{k+2+\delta}(\Omega)$, then by the method of continuity we could get the solvability of the equation $Lu = f$ on the sole basis of estimate (6.3.1).

6.4. The regularizer

Take the bounded domain $\Omega \in C^{k+2+\delta}$ and take $n, x_1, ..., x_n, \zeta^i$ from Sec. 6.3. In Sec. 6.1 we have constructed the operators $S_z^{x_0}$ for any $z > 0$ and $x_0 \in \partial\Omega$. Let $S_z^i = S_z^{x_i}, i = 1, 2, ..., n$, and let $S_z^0 = \mathcal{R}_z$, where the last operator is taken from Remark 4.3.5.

A naive idea to solve $zu - Lu = f$ in Ω is to define

$$u = \begin{cases} S_z^0 f & \text{in} \quad \Omega \setminus \bigcup_i B_{\rho_0/2}(x_i), \\ S_z^i f & \text{in} \quad B_{\rho_0/2}(x_i) \setminus \bigcup_{j \leq i-1} B_{\rho_0/2}(x_j). \end{cases}$$

Unfortunately this function u is just discontinuous, although by Exercise 6.2.4 it satisfies the equation in Ω apart from boundaries of the pieces where it is defined by different formulas.

Therefore, we slightly modify this idea.

LEMMA 6.4.1. *Let $z > 0$ and $u \in C_0^{2+\delta}(\Omega)$ and $zu - Lu = f$. Then in Ω*

$$u = \sum_{i \leq n} \zeta^i S_z^i(\zeta^i f - L^i u), \tag{6.4.1}$$

where

$$L^k u := u(a^{ij}\zeta^k_{x^i x^j} + b^i \zeta^k_{x^i}) + 2a^{ij}\zeta^k_{x^i} u_{x^j} \quad (= L(\zeta^k u) - \zeta^k Lu).$$

Proof. It should be said that in (6.4.1) we have terms $S_z^i(...)$ which, formally, are *not* defined everywhere in Ω. But we multiply them by the functions ζ^i which vanish outside the set where $S_z^i(...)$ is defined, and we define these products to be zero there.

Next, clearly $\zeta^0 u \in C^{2+\delta}(\mathbb{R}^d)$ and by the definition of S_z^0 we have

$$\zeta^0 u = S_z^0(z(\zeta^0 u) - L(\zeta^0 u)).$$

Hence by Theorem 6.2.2 (i)

$$u = \sum_{i \leq n} \zeta^i(\zeta^i u) = \sum_{i \leq n} \zeta^i S_z^i(z(\zeta^i u) - L(\zeta^i u)) = \sum_{i \leq n} \zeta^i S_z^i(\zeta^i f - L^i u).$$

The lemma is proved.

DEFINITION 6.4.2. By a *regularizer* of the operator $z - L$ we mean the operator

$$\boldsymbol{R}_z f = \sum_{i \leq n} \zeta^i S_z^i(\zeta^i f).$$

From (6.4.1) we see that the regularizer almost gives us the inverse to $z - L$. The only operator which interferes is

$$u \to \sum_{i \le n} \zeta^i S^i_z(-L^i u),$$

but as we will see this operator is "weaker" than the regularizer.

6.5. The existence theorems

Take the bounded domain $\Omega \in C^{k+2+\delta}$, and take $n, x_1, ..., x_n$ from Sec. 6.3 and the regularizer from Sec. 6.4. Instead of solving the equation $zu - Lu = f$ in $C_0^{k+2+\delta}(\Omega)$ we want to solve equation (6.4.1). First of all we need to know that any solution of the latter equation is indeed a solution of $zu - Lu = f$.

LEMMA 6.5.1. *If $z > 0$, $f \in C^{k+\delta}(\Omega)$, and $u \in C^{1+\delta}(\Omega)$ is a solution of (6.4.1), then $u \in C_0^{k+2+\delta}(\Omega)$. Furthermore, there exists a constant $z_0 \ge 1$, depending only on d, K, κ, k, δ, K_0, ρ_0, and d_Ω, such that if, in addition, $z \ge z_0$, then $zu - Lu = f$ in Ω.*

Proof. By previous results (see Remarks 5.4.4 and 4.3.5 and Lemma 6.1.8) from equality (6.4.1) and inclusions $\zeta^i f - L^i u \in C^\delta(\Omega)$ it follows that $u \in C_0^{2+\delta}(\Omega)$. In turn, if $k \ge 1$, then we have $\zeta^i f - L^i u \in C^{1+\delta}(\Omega)$, and by the same results $u \in C_0^{3+\delta}(\Omega)$. In this manner we see that $u \in C_0^{k+2+\delta}(\Omega)$ indeed.

Next, denote $g = zu - Lu$. Then by Lemma 6.4.1 equality (6.4.1) holds with g in place of f, and to finish the proof we need only to show that if z is large and $h \in C^\delta(\Omega)$ and $\boldsymbol{R}_z h = 0$ in Ω, then $h = 0$ in Ω.

By Exercise 6.2.4 we have $(z - L)S^i_z(\zeta^i h) = \zeta^i h$ on the set in Ω where $\zeta^i \ne 0$. Therefore, from $\boldsymbol{R}_z h = 0$ we find

$$0 = (z - L)\boldsymbol{R}_z h = \sum_{i=0}^{n} \zeta^i (z - L) S^i_z(\zeta^i h) - \sum_{i=0}^{n} L^i S^i_z(\zeta^i h)$$

$$= \sum_{i=0}^{n} (\zeta^i)^2 h - \sum_{i=0}^{n} L^i S^i_z(\zeta^i h) = h - T_z h,$$

where

$$T_z h := \sum_{i=0}^{n} L^i S^i_z(\zeta^i h).$$

To finish the proof it suffices to show that for z large the operator T_z is a contraction in $C^\delta(\Omega)$.

By Theorem 6.2.2 and Remark 4.3.5 we have

$$|T_z h|_{\delta;\Omega} \le N \sum_{i=1}^{n} |S^i_z(\zeta^i h)|_{1+\delta; B_{\rho_0}(x_i)\cap\Omega} + N|S^0_z(\zeta^0 h)|_{1+\delta}$$

$$\le N z^{(\delta-1)/2} \sum_{i=1}^{n} |\zeta^i h|_{\delta; B_{\rho_0}(x_i)\cap\Omega} + N z^{(\delta-1)/2} |\zeta^0 h|_\delta \le N z^{(\delta-1)/2} |h|_{\delta;\Omega}.$$

Since the last constant N does not depend on z, the lemma is proved.

LEMMA 6.5.2. *There is a constant $z_0 = z_0(d, K, \kappa, \delta, K_0, \rho_0, d_\Omega) \geq 1$ such that for any $z \geq z_0$ and any $f \in C^\delta(\Omega)$ there exists a unique solution $u \in C^{1+\delta}(\Omega)$ of equation (6.4.1).*

To prove this lemma it suffices to show that for z large the operator

$$u \to \sum_{i=0}^{n} \zeta^i S_z^i(L^i u)$$

is a contraction in $C^{1+\delta}(\Omega)$. This is made exactly as in the end of the previous proof.

THEOREM 6.5.3. *For any $f \in C^{k+\delta}(\Omega)$ and $g \in C^{k+2+\delta}(\Omega)$ there exists a unique function $u \in C^{k+2+\delta}(\bar{\Omega})$ satisfying the equation $Lu = f$ in Ω and equal to g on $\partial\Omega$.*

Proof. By considering $u-g$ instead of u, one reduces the general case to the case $g = 0$. Then Lemmas 6.5.2 and 6.5.1 show that for z large enough the equation $Lu - zu = f$ is solvable in $C_0^{k+2+\delta}(\Omega)$. One can fix an appropriate z and then apply the method of continuity, bearing in mind that Theorem 6.3.2 provides the necessary a priori estimate. The theorem is proved.

COROLLARY 6.5.4. *If $\partial\Omega$ is infinitely differentiable and f, g and the coefficients of L are infinitely differentiable in $\mathbb{R}^d$, then the solution u is infinitely differentiable in $\bar{\Omega}$.*

REMARK 6.5.5. The above proof of solvability of elliptic equations in domains uses partitions of unity which makes it possible to carry out the same proof in the case of elliptic equations on manifolds with or without boundary.

EXERCISE 6.5.6. Let $L = \Delta$ and $\Omega \in C^{2+\delta}$. By using Theorem 2.3.1 prove existence of the Green's function $G(x, y)$ and the Poisson kernel for Ω.

EXERCISE 6.5.7. In the situation described in Exercise 6.5.6 prove that $G(x, y) = G(y, x)$ for any $x \neq y$, $x, y \in \Omega$.

6.6. Finite–difference approximations of elliptic operators

One might feel uncomfortable seeing only existence and uniqueness theorems. Indeed, the most important thing in real applications is to find the solution. Of course, we cannot hope to find explicit formulas for solutions for general elliptic operators. By the way, even if one manages to find such a formula for a particular problem and the formula is complicated, one has to find numerical methods to be able to use the formula for real computations.

In this and the following section we show that the general theory can indeed be applied to finding numerical approximations of solutions with any given accuracy.

Take a bounded domain $\Omega \in C^{2+\delta}$ and let

$$L = a^{ij}(x)\frac{\partial^2}{\partial x^i \partial x^j} + b^i(x)\frac{\partial}{\partial x^i} + c(x)$$

be an elliptic operator with constant of ellipticity κ and with $|a, b, c|_\delta \leq K$, where κ, K are some strictly positive constants. Take a number $h \in (0, 1]$ and denote

$$Z_h^d = \{x : x = h\sum_{i=1}^d e_i n_i, n_i = 0, \pm1, \pm2, ...\}$$

the uniform h–grid on $\mathbb{R}^d$. Also let $\Omega_h = \Omega \cap Z_h^d$ and denote by Ω_h^o the "interior" of Ω_h that is the set of all points x in Ω_h such that dist $(x, \partial\Omega) \geq h$. The set $\partial\Omega_h = \Omega_h \setminus \Omega_h^o$ is interpreted as the boundary of the discretized domain Ω_h.

Assume that for any $h \in (0,1], x \in \Omega_h^o, y \in \Omega_h$ we are given some numbers $p_h(x,y)$ and denote

$$L_h u(x) = \sum_{y\in\Omega_h} p_h(x,y)u(y). \tag{6.6.1}$$

We want the operators L_h to "behave" like L and to approximate L as $h \downarrow 0$. To this end we make the following assumptions.

ASSUMPTION 6.6.1 (maximum principle). If u is a function defined on Ω_h and for a point $x_0 \in \Omega_h^o$ we have $u(x_0) = \max_{\Omega_h} u(x) > 0$, then $L_h u(x_0) \leq 0$.

ASSUMPTION 6.6.2. The operators L_h approximate L. More precisely, for any $u \in C^{2+\delta}(\Omega)$ and any $x \in \Omega_h^o$ we have

$$|Lu(x) - L_h u(x)| \leq Kh^\delta |u|_{2+\delta;\Omega}.$$

Let us show how the operators L_h can be constructed in a particular case.

EXAMPLE 6.6.3. For simplicity we consider only operators without mixed derivatives. In other words we assume that $a^{ij} \equiv 0$ for $i \neq j$. Then define

$$L_h u(x) = \sum_{i=1}^d a^{ii}(x)h^{-2}[u(x+he_i) - 2u(x) + u(x-he_i)]$$

$$+\sum_{i=1}^d |b^i(x)|h^{-1}[u(x+he_i\text{sign}\, b^i(x)) - u(x)] + c(x)u(x).$$

This operator certainly has form (6.6.1). Moreover, owing to the inequalities $a^{ii}, |b^i|, h \geq 0$ and $c \leq 0$, the operator L_h satisfies the maximum principle. Finally, for $u \in C^{2+\delta}(\Omega)$ and $x \in \Omega_h^o$ we have

$$|L_h u(x) - Lu(x)| \leq N\sum_{i=1}^d |h^{-2}[u(x+he_i) - 2u(x) + u(x-he_i)] - u_{x^ix^i}(x)|$$

$$+ N\sum_{i=1}^d |[u(x+he_i\text{sign}\, b^i(x)) - u(x)]h^{-1}\text{sign}\, b^i(x) - u_{x^i}(x)|$$

$$\leq Nh^\delta[u]_{2+\delta;\Omega} + Nh[u]_{2;\Omega},$$

so that Assumption 6.6.2 is satisfied too.

If $b^i \equiv 0$, then for $u \in C^{3+\delta}(\Omega)$ the expression $L_h u$ approximates Lu with *better* accuracy. Indeed, in that case for $x \in \Omega_h^o$ by Taylor's formula

$$h^{-2}[u(x+he_i)-2u(x)+u(x-he_i)] = u_{x^ix^i}(x)+\frac{1}{6}h[u_{x^ix^ix^i}(y)-u_{x^ix^ix^i}(z)],$$

where y,z lie on the straight segment $[x-he_i,x+he_i]\subset\Omega$. Hence

$$|L_hu(x)-Lu(x)|\le Nh^{1+\delta}|u|_{3+\delta;\Omega}.$$

EXERCISE* 6.6.4. Prove that under Assumption 6.6.1 one has $p_h(x,x)\le 0$ and $p_h(x,y)\ge 0$ for $x\ne y$.

6.7. Convergence of numerical approximations

We continue to study the situation described in the previous section, keeping the notation and assumptions from there.

LEMMA 6.7.1. *There is a constant $h_0>0$ depending only on κ,K,δ,d and the diameter of Ω such that for $h\in(0,h_0]$ for any bounded functions f,g the system of linear equations*

$$L_hu(x)+f(x)=0\quad\forall x\in\Omega_h^o,\quad u(x)=g(x)\quad\forall x\in\partial\Omega_h \tag{6.7.1}$$

has a unique solution $u_h(x),x\in\Omega_h$. In addition

$$\max_{\Omega_h}(u_h(x))_+\le N\max_{\Omega_h^o}f_+(x)+\max_{\partial\Omega_h}g_+(x),$$

$$\max_{\Omega_h}(u_h(x))_-\le N\max_{\Omega_h^o}f_-(x)+\max_{\partial\Omega_h}g_-(x),$$

$$\max_{\Omega_h}|u_h(x)|\le N\max_{\Omega_h^o}|f(x)|+\max_{\partial\Omega_h}|g(x)|, \tag{6.7.2}$$

where the constant N depends only on κ,K,d and the diameter of Ω.

Proof. Let n be the number of points in Ω_h. Then it is seen that the linear system (6.7.1) is a system of n equations about n variables $u_h(x),x\in\Omega_h$. Therefore, to prove the first assertion we only need prove uniqueness of the trivial solution for $f\equiv g\equiv 0$. Of course, this uniqueness follows at once from (6.7.2).

As far as (6.7.2) is concerned, it suffices only to prove the first estimate, in the proof of which without loss of generality we assume that $0\in\Omega$. Then we take the function v_0 from Lemma 6.1.1 with R defined as the diameter of Ω. Observe that $Lv_0\le -1$ in Ω, so that by Assumption 6.6.2 we can choose h_0 to have $L_hv_0\le -1/2$ for any $h\in(0,h_0]$ and any $x\in\Omega_h^o$.

Now take a solution u_h of (6.7.1) and consider $w=u_h-2(F+\varepsilon)v_0-G$, where $F=\max_{\Omega_h^o}f_+$, $G=\max_{\partial\Omega_h}g_+$ and the constant $\varepsilon>0$. If we prove that for any ε we have $w\le 0$ in Ω_h, the first estimate in (6.7.2) will obviously follow.

Assume that $w>0$ at some points and define x_0 as a point in Ω_h where w takes its maximum value $w(x_0)>0$. Since $u_h=g$ and $v_0\ge 0$ on $\partial\Omega_h$, we have $x_0\in\Omega_h^o$. By Assumption 6.6.1 we obtain $L_hG\le 0$ and

$$0\ge L_hw(x_0)=-f(x_0)-2(F+\varepsilon)L_hv_0(x_0)-L_hG\ge -f(x_0)+F+\varepsilon\ge\varepsilon>0.$$

We get the desired contradiction, and the lemma is proved.

THEOREM 6.7.2. *Let $f \in C^{\delta}(\Omega), g \in C^{2+\delta}(\Omega)$, $h \in (0, h_0]$ and denote by u_h the corresponding solution of (6.7.1). Also let $u \in C^{2+\delta}(\Omega)$ be a unique solution of the problem: $Lu + f = 0$ in Ω, $u = g$ on $\partial\Omega$. Then*

$$|u - u_h|_{0;\Omega_h} \le Nh^{\delta}\Big(|f|_{\delta;\Omega} + |g|_{2+\delta;\Omega}\Big), \tag{6.7.3}$$

where the constant N depends only on d, K, δ, κ, the diameter of Ω and the constants ρ_0, K_0 from Definition 6.1.6.

Proof. For $x \in \Omega_h^o$ we have

$$\begin{aligned}|L_h(u_h - u)(x)| =&|f(x) + L_h u(x)| = |(L - L_h)u(x)| \\ \le&Kh^{\delta}|u|_{2+\delta;\Omega} \le Nh^{\delta}\Big(|f|_{\delta;\Omega} + |g|_{2+\delta;\Omega}\Big).\end{aligned} \tag{6.7.4}$$

Also, if $x \in \partial\Omega_h$, then the distance from x to $\partial\Omega$ is less than h, so that there is a $y \in \partial\Omega$ satisfying $|x - y| \le h$ and

$$\begin{aligned}|(u_h - u)(x)| =&|g(x) - u(x)| \le |g(x) - g(y)| + |u(y) - u(x)| \\ \le&Nh\Big(|g|_{2+\delta;\Omega} + |u|_{2+\delta;\Omega}\Big).\end{aligned} \tag{6.7.5}$$

It remains only to apply Lemma 6.7.1. The theorem is proved.

REMARK 6.7.3. We see that our estimate (6.7.3) of the error of approximating u by u_h comes from two sources: from the right-hand sides of (6.7.4) and (6.7.5). The first expression is an estimate of the error of approximating the operator L by L_h, and the second one arises since $\partial\Omega_h$ does not lie on the boundary of Ω. The second one can be made zero if one manages to place all points of $\partial\Omega_h$ on $\partial\Omega$. This is always possible if one is free to choose not necessarily rectangular mesh. In fact, we have not used at all any specific information about the structure of Ω_h. It is also useful to bear in mind that sometimes, as in Example 6.6.3, one can approximate L with better accuracy than in Assumption 6.6.2.

EXERCISE 6.7.4. Theorem 6.7.2 shows what may happen only in the worst possible case. Usually in practice the convergence is faster. In connection with this take the equation $u'' = 1$ on $(-1, 1)$ with the boundary condition $u(\pm 1) = 0$ and show that $u(x) = u_h(x)$ (with no error) at all points $k/n \in (-1, 1)$, where $h = 1/n$ and u_h is the solution of $n^2\Big(u((k+1)/n) - 2u(k/n) + u((k-1)/n)\Big) = 1$, $k = 0, \pm 1, ..., \pm(n-1)$, $u(\pm 1) = 0$.

6.8. Hints to exercises

6.1.5. By a direct computation show that there exists $\nu < 0$ such that in Ω

$$L\Big(\frac{(rR)^{\nu} - |x|^{\nu}}{(rR)^{\nu} - R^{\nu}}\Big) \le (a^{ij}D_iD_j + b^iD_i)\Big(\frac{(rR)^{\nu} - |x|^{\nu}}{(rR)^{\nu} - R^{\nu}}\Big) \le 0.$$

6.5.7. Observe that by Exercise 2.6.6 the function $G(x, y)$ is continuous in (x, y) for $x \ne y$, $x, y \in \Omega$. Then take $\psi_1, \psi_2 \in C_0^{\infty}(\mathbb{R}^d)$ such that the supports of ψ_i lie in Ω and are disjoint. Denote by u_i solutions of $\Delta u = \psi_i$ with zero boundary condition and use

$$\int_\Omega u_1 \Delta u_2(x)\, dx = \int_\Omega u_2(x) \Delta u_1(x)\, dx, \quad u_i(x) = \int_\Omega G(x,y)\psi_i(y)\, dy,$$

$$\int_\Omega \Big(\int_\Omega G(x,y)\psi_1(y)\, dy \Big) \psi_2(x)\, dx = \int_\Omega \psi_1(x) \Big(\int_\Omega G(x,y)\psi_2(y)\, dy \Big)\, dx.$$

CHAPTER 7

Elliptic Equations in Non-Smooth Domains

In this chapter, except for a part of Sec. 7.1, we are considering the Dirichlet problem for second–order elliptic equations with real coefficients. We show that this equations are solvable even if boundaries of domains are not very regular and solutions are continuous up to the boundary and have necessary derivatives continuous inside domains.

7.1. Interior a priori estimates

The results of this section will be later used only for second–order elliptic operators, but since the proofs do not depend on the order, we choose to treat the case of general elliptic operators.

Let L be an mth order elliptic operator with (complex) coefficients a^α, $|\alpha| \leq m$, and constant of ellipticity $\kappa > 0$. Take an integer $k \geq 0$ and a constant K and assume that $|a^\alpha|_{k+\delta} \leq K$.

THEOREM 7.1.1. *Let $R > 0$; then there is a constant N depending only on $R, \kappa, K, k, \delta, d$ such that for any $\lambda \in [-1, 1]$ and $u \in C^{k+m+\delta}(B_{2R})$ we have*

$$|u|_{k+m+\delta;B_R} \leq N\Big(|L_\lambda u|_{k+\delta;B_{2R}} + |u|_{0;B_{2R}}\Big). \tag{7.1.1}$$

Proof. Dilations show that without loss of generality we can assume $R = 1$. As in Exercise 4.2.3 one proves that there is a constant N depending only on κ, K, k, δ, d such that for any $\lambda \in [-1, 1]$ and $u \in C^{k+m+\delta}(\mathbb{R}^d)$ we have

$$|u|_{k+m+\delta} \leq N\Big(|L_\lambda u|_{k+\delta} + |u|_0\Big). \tag{7.1.2}$$

Next, let $R_n = \sum_{j=0}^n 2^{-j}$, $n = 0, 1, 2, \dots$. We need some functions $\zeta_n \in C_0^\infty(\mathbb{R}^d)$ such that $\zeta_n(x) = 1$ in B_{R_n}, $\zeta_n(x) = 0$ outside $B_{R_{n+1}}$ and

$$|\zeta_n|_{k+m+\delta} \leq N 2^{n(k+m+1)} = N\rho^{-n},$$

where $\rho = 2^{-k-m-1} < 1$ and $N = N(d, k, m)$. To construct them take an infinitely differentiable function $h(t)$, $t \in (-\infty, \infty)$, such that $h(t) = 1$ for $t \leq 1$, $h(t) = 0$ for $t \geq 2$ and $0 \leq h \leq 1$. After this define

$$\zeta_n(x) = h(2^{n+1}(|x| - R_n + 2^{-(n+1)})).$$

Now we put $u\zeta_n$ in (7.1.2) to get

$$
\begin{aligned}
|u|_{k+m+\delta;B_{R_n}} \le & N\Big(|L_\lambda(u\zeta_n)|_{k+\delta;B_{R_{n+1}}} + |u|_{0;B_{R_{n+1}}}\Big) \\
\le & N\Big(|L_\lambda u|_{k+\delta;B_{R_{n+1}}} + \rho^{-n}|u|_{k+m-1+\delta;B_{R_{n+1}}} + |u|_{0;B_{R_{n+1}}}\Big) \qquad (7.1.3) \\
\le & N|L_\lambda u|_{k+\delta;B_{R_{n+1}}} + N_1\rho^{-n}|u|_{k+m-1+\delta;B_{R_{n+1}}}.
\end{aligned}
$$

By interpolation inequalities (see Lemma 6.3.1) for $\gamma = (k+m)/(k+m+\delta)$ and $\varepsilon \in (0,1)$

$$
\begin{aligned}
|u|_{k+m-1+\delta;B_{R_{n+1}}} \le & \, N\Big(|u|^{\gamma}_{k+m+\delta;B_{R_{n+1}}}|u|^{1-\gamma}_{0;B_{R_{n+1}}} + |u|_{0;B_{R_{n+1}}}\Big) \\
\le & \, \varepsilon^{1/\gamma}|u|_{k+m+\delta;B_{R_{n+1}}} + N\varepsilon^{-1/(1-\gamma)}|u|_{0;B_2}.
\end{aligned}
$$

Therefore, coming back to (7.1.3) and denoting $\xi = N_1\rho^{-n}\varepsilon^{1/\gamma}$, for any $\xi \in (0,1)$ we get

$$
|u|_{k+m+\delta;B_{R_n}} \le N|L_\lambda u|_{k+\delta;B_2} + \xi|u|_{k+m+\delta;B_{R_{n+1}}} + N\xi^{-\gamma/(1-\gamma)}\rho^{-n/(1-\gamma)}|u|_{0;B_2}.
$$

Further, define $\xi = \rho^{1/(1-\gamma)}/2$. Then $\xi^{-\gamma/(1-\gamma)}\rho^{-n/(1-\gamma)} = N\rho^{-n/(1-\gamma)} = N(2\xi)^{-n}$ and

$$
|u|_{k+m+\delta;B_{R_n}} \le N|L_\lambda u|_{k+\delta;B_2} + \xi|u|_{k+m+\delta;B_{R_{n+1}}} + N(2\xi)^{-n}|u|_{0;B_2},
$$

$$
\xi^n|u|_{k+m+\delta;B_{R_n}} \le N\xi^n|L_\lambda u|_{k+\delta;B_2} + \xi^{n+1}|u|_{k+m+\delta;B_{R_{n+1}}} + N2^{-n}|u|_{0;B_2},
$$

$$
\sum_{n=0}^{\infty}\xi^n|u|_{k+m+\delta;B_{R_n}} \le N|L_\lambda u|_{k+\delta;B_2}\sum_{n=0}^{\infty}\xi^n + \sum_{n=1}^{\infty}\xi^n|u|_{k+m+\delta;B_{R_n}} + N|u|_{0;B_2}.
$$

To prove (7.1.1) it remains only to collect like terms in the last inequality after noticing that its left-hand side is finite, since $\xi < 1$ and $|u|_{k+m+\delta;B_{R_n}} \le |u|_{k+m+\delta;B_2} < \infty$. The theorem is proved.

Theorem 7.1.1 is a quantitative form of the following interior regularity result, which does not require Theorem 7.1.1.

THEOREM 7.1.2. *Let Ω be a domain in $\mathbb{R}^d$ and $u \in C^{m+\delta}(\Omega)$. Assume that $L_\lambda u \in C^{k+\delta}(\Omega)$ for some λ. Then $u \in C^{k+m+\delta}_{loc}(\Omega)$.*

Proof. At first assume that λ is so large that the equation $L_\lambda v = f$ in $\mathbb{R}^d$ is uniquely solvable in $C^{m+\delta}(\mathbb{R}^d)$ for any $f \in C^\delta(\mathbb{R}^d)$, and if $f \in C^{k+\delta}(\mathbb{R}^d)$, then $v \in C^{k+m+\delta}(\mathbb{R}^d)$. Then take any $\zeta \in C_0^\infty(\Omega)$ and notice that

$$
L_\lambda(\zeta u)(x) = \zeta(x)Lu(x) + \sum_{|\alpha|\le m-1} b^\alpha(x)D^\alpha u(x) =: f,
$$

where $b^\alpha = 0$ near the boundary of Ω, and $b^\alpha \in C^{k+\delta}(\mathbb{R}^d)$. It follows that if $k \ge 1$, then $f \in C^{1+\delta}(\mathbb{R}^d)$. By the assumption the equation $L_\lambda v = f$ has a solution which is unique in $C^{m+\delta}(\mathbb{R}^d)$, and in addition this solution belongs to $C^{1+m+\delta}(\mathbb{R}^d)$. Since

ζu is also a $C^{m+\delta}(\mathbb{R}^d)$–solution, we obtain that $\zeta u \in C^{1+m+\delta}(\mathbb{R}^d)$. This implies that $u \in C^{1+m+\delta}_{loc}(\Omega)$.

If $k \geq 2$, then $k \geq 1$ and $u \in C^{1+m+\delta}_{loc}(\Omega)$ and $f \in C^{2+\delta}(\mathbb{R}^d)$. As above we now get $\zeta u \in C^{2+m+\delta}(\mathbb{R}^d)$, that is, $u \in C^{2+m+\delta}_{loc}(\Omega)$, and so on.

Next we claim that if the assertion of the theorem is true for at least one λ, it is also true for any λ. To show this it suffices as in the proof of Theorem 4.2.1 to notice that $L_\lambda u = L_\mu u + Mu$, where M is an $(m-1)$st order operator with $C^{k+\delta}(\mathbb{R}^d)$ coefficients. The theorem is proved.

REMARK 7.1.3. For Theorem 7.1.2 to be true we do not need the coefficients of L to be of class $C^{k+\delta}$ outside Ω.

Indeed, without any loss of generality we could take balls B_R instead of general domains Ω, assuming that the coefficients are of class $C^{k+\delta}(B_{2R})$. In that case we can always change the coefficients outside B_R, getting functions of class $C^{k+\delta}(\mathbb{R}^d)$ and preserving the ellipticity of L. If $\Omega = B_1$, this can be done, for instance, by the formula $a^\alpha(x) = a^\alpha(\pi(x))$, where $\pi(x)$ is any C_0^∞-function with range belonging to B_2 and such that $\pi(x) = x$ if $|x| \leq 1$.

COROLLARY 7.1.4. *Let Ω be a domain in $\mathbb{R}^d$ and $u \in C^{m+\delta}(\Omega)$. Assume that the coefficients a^α and the function $L_\lambda u$ are infinitely differentiable in Ω for some λ. Then u is infinitely differentiable in Ω.*

EXERCISE 7.1.5. Let L be a second-order elliptic operator with real coefficients and constant of ellipticity $\kappa > 0$. Assume that $|a, b, c|_{k+\delta} \leq K$ and $\Omega \in C^{k+2+\delta}$. Prove that for *any* $R > 0$

(i) if $u \in C^{k+2+\delta}(B_{2R} \cap \Omega)$ and $u = 0$ on $B_{2R} \cap \partial\Omega$, then

$$|u|_{k+2+\delta; B_R \cap \Omega} \leq N\Big(|Lu|_{k+\delta; B_{2R}\cap\Omega} + |u|_{0; B_{2R}\cap\Omega}\Big),$$

where N is independent of u;

(ii) if $u \in C^{2+\delta}(B_{2R} \cap \Omega)$ and $u = 0$ on $B_{2R} \cap \partial\Omega$ and $Lu \in C^{k+\delta}(B_{2R} \cap \Omega)$, then $u \in C^{k+2+\delta}(B_R \cap \Omega)$.

EXERCISE 7.1.6. Let L be a second-order nonuniformly elliptic operator with real coefficients, which means that $L = a^{ij}D_iD_j + b^iD_i + c$, where a, b, c are real, $a^{ij}(x) = a^{ji}(x)$, $c \leq 0$ and at any point $x \in \mathbb{R}^d$ the matrix $a(x)$ is strictly positive. Assume that $|a, b, c, f|_{k+\delta; B_R} < \infty$ for any $R > 0$. Also assume that $|f| \leq -c$ in $\mathbb{R}^d$. Prove that there exists a solution $u \in C^{k+2+\delta}_{loc}(\mathbb{R}^d)$ of the equation $Lu + f = 0$. Observe that this u need not be unique, as follows from the remark after Exercise 2.9.5.

7.2. Generalized solutions of the Dirichlet problem with zero boundary condition

From now on in this chapter we fix an elliptic second–order operator L with a constant of ellipticity $\kappa > 0$ and with coefficients $a, b, c \in C^\delta(\mathbb{R}^d)$. If Ω is a bounded domain of class $C^{2+\delta}$ and $f \in C^\delta(\Omega)$, then by $\mathcal{R}(\Omega)f$ we denote the solution $u \in C^{2+\delta}(\Omega)$ to the following Dirichlet problem

$$Lu + f = 0 \quad \text{in} \quad \Omega, \quad u = 0 \quad \text{on} \quad \partial\Omega.$$

LEMMA 7.2.1. *Let $\Omega \in C^{2+\delta}$. (i) If $f, g \in C^{\delta}(\Omega)$ and $f \geq g$, then $\mathcal{R}(\Omega)f \geq \mathcal{R}(\Omega)g$. Moreover, $|\mathcal{R}(\Omega)f| \leq \mathcal{R}(\Omega)|f|$.*

(ii) If $\Omega_1, \Omega_2 \in C^{2+\delta}$ and $\Omega_1 \subset \Omega_2$ and $f, g \in C^{\delta}(\Omega_2)$ and $f \geq g_+$, then $\mathcal{R}(\Omega_1)g \leq \mathcal{R}(\Omega_2)f$ in Ω_1.

(iii) There exists a constant N depending only on κ, d and maxima of $|a^{ij}|$, $|b^i|$, $|c|$ and the diameter of Ω such that $|\mathcal{R}(\Omega)f|_{0;\Omega} \leq N|f|_{0;\Omega}$.

(iv) Let Ω_n be a sequence of domains of class $C^{2+\delta}$ such that $\Omega_n \subset \Omega_{n+1} \subset \Omega$ and $\Omega = \cup_n \Omega_n$. Let $f \in C^{\delta}(\Omega)$. Then for any $x \in \Omega$

$$\mathcal{R}(\Omega)f(x) = \lim_{n\to\infty} \mathcal{R}(\Omega_n)f(x).$$

Proof. All these assertions are easy consequences of the maximum principle. Indeed, in (i) we have $-L[\mathcal{R}(\Omega)f - \mathcal{R}(\Omega)g] = f - g \geq 0$ and $\mathcal{R}(\Omega)f - \mathcal{R}(\Omega)g = 0$ on $\partial\Omega$. Hence $\mathcal{R}(\Omega)f - \mathcal{R}(\Omega)g \geq 0$. From $-|f| \leq f \leq |f|$ we have $-\mathcal{R}(\Omega)|f| \leq \mathcal{R}(\Omega)f \leq \mathcal{R}(\Omega)|f|$, so that $|\mathcal{R}(\Omega)f| \leq \mathcal{R}(\Omega)|f|$.

(ii) Since $g_+ \geq 0$, also $f \geq 0$ and $\mathcal{R}(\Omega_2)f \geq \mathcal{R}(\Omega)0 = 0$. Therefore, $\mathcal{R}(\Omega_2)f - \mathcal{R}(\Omega_1)g \geq 0$ on $\partial\Omega_1$. By adding that $-L[\mathcal{R}(\Omega_2)f - \mathcal{R}(\Omega_1)g] = f - g \geq 0$ in Ω_1, we get $\mathcal{R}(\Omega_2)f - \mathcal{R}(\Omega_1)g \geq 0$ in Ω_1.

Assertion (iii) is actually proved in the proof of Theorem 6.3.2. To prove (iv) define $u = \mathcal{R}(\Omega)f$, $u_n = \mathcal{R}(\Omega_n)f$. We have to show that $u_n(x) \to u(x)$ at any $x \in \Omega$.

Define $\varepsilon_n = \sup\{|u|; \partial\Omega_n\}$. Since $\Omega_n \subset \Omega_{n+1}$ and $\Omega = \cup_n \Omega_n$, for any $\gamma > 0$ there is a number $n(\gamma)$ such that starting with this $n(\gamma)$ the boundaries of Ω_n lie in the γ-neighborhood of $\partial\Omega$. Also, u is continuous in $\bar{\Omega}$ and $u = 0$ on $\partial\Omega$, which implies that $\varepsilon_n \to 0$ as $n \to \infty$. From the definition of ε_n it follows that $L(u - u_n - \varepsilon_n) = -c\varepsilon_n \geq 0$ in Ω_n and $u - u_n - \varepsilon_n = u - \varepsilon_n \leq 0$ on $\partial\Omega_n$. By the maximum principle $u - u_n - \varepsilon_n \leq 0$ and $u - u_n \leq \varepsilon_n$ in Ω_n. Similarly, $u_n - u \leq \varepsilon_n$ in Ω_n, so that $|u_n - u| \leq \varepsilon_n$ in Ω_n, which gives that $u_n \to u$ in Ω indeed. The lemma is proved.

DEFINITION 7.2.2. Let Ω be a bounded domain, Ω_n a sequence of subdomains of class $C^{2+\delta}$ such that $\Omega_n \subset \Omega_{n+1} \subset \bar{\Omega}_{n+1} \subset \Omega$, $\Omega = \cup_n \Omega_n$. For $f \in C^{\delta}(\Omega)$ and $x \in \Omega$ define

$$\mathcal{R}(\Omega)f(x) = \lim_{n\to\infty} \mathcal{R}(\Omega_n)f(x). \tag{7.2.1}$$

Of course, we need to show that this definition makes sense and introduces the object $\mathcal{R}(\Omega)$, which in the case of smooth Ω coincides with the one introduced before. By the way, the latter follows from Lemma 7.2.1 (iv).

To show that the definition makes sense, first note that by Lemma 7.2.1, for $f \geq 0$ the sequence $\mathcal{R}(\Omega_n)f(x)$ increases and is bounded. Therefore, in this case the limit on the right in (7.2.1) exists and is finite. For general f we use $f = f_+ - f_-$ and $\mathcal{R}(\Omega_n)f = \mathcal{R}(\Omega_n)f_+ - \mathcal{R}(\Omega_n)f_-$.

Next we claim that $\mathcal{R}(\Omega)$ is independent of the choice of subdomains Ω_n. To prove this it suffices to take any other sequence of domains Ω'_n satisfying the conditions of Definition 7.2.2, then construct an alternating sequence so that domains with odd numbers are coming from $\{\Omega_n\}$ and with even ones from $\{\Omega'_n\}$ (remember that $\bar{\Omega}_n \subset \Omega$ and the sets $\bar{\Omega}_n$ are compact), and finally upon noticing that the limit in (7.2.1) exists along the alternating sequence as well, conclude that the limits for Ω_n and Ω'_n coincide.

After we have justified Definition 7.2.2, an application of Theorem 7.1.1 on interior estimates yields that if $\Omega \in C^{2+\delta}$, $f \in C^{\delta}(\Omega)$ and a domain $\Omega' \subset \bar{\Omega}' \subset \Omega$, then

$$|\mathcal{R}(\Omega)f|_{2+\delta;\Omega'} \le N(|f|_{\delta;\Omega} + |\mathcal{R}(\Omega)f|_{0;\Omega}) \le N|f|_{\delta;\Omega},$$

where the constant N depends only on the distance between boundaries of Ω' and Ω and d, K, κ, δ and the diameter of Ω. An obvious passage to the limit from smooth Ω to arbitrary bounded ones gives the following result.

THEOREM 7.2.3. *If Ω is a bounded domain and $f \in C^{\delta}(\Omega)$, then $\mathcal{R}(\Omega)f \in C^{2+\delta}_{loc}(\Omega)$ and $L\mathcal{R}(\Omega)f + f = 0$ in Ω.*

Further, the operators $\mathcal{R}(\Omega)$ obviously possess properties (i), (ii) and (iii) from Lemma 7.2.1 for any bounded domains Ω, Ω_1, Ω_2. It is known that any continuous function f given on a compact set, say $\bar{\Omega}$, can be extended to $\mathbb{R}^d$ to a continuous and bounded function (Uryson's theorem). This implies that such an f can be uniformly approximated in $\bar{\Omega}$ by polynomials and along with property (iii) shows that the following definition makes sense.

DEFINITION 7.2.4. For any bounded domain Ω and any function $f \in C(\bar{\Omega})$ take any sequence of functions (for instance, polynomials) $f_n \in C^1(\bar{\Omega})$ such that $|f - f_n|_{0;\Omega} \to 0$, and for $x \in \Omega$ set

$$\mathcal{R}(\Omega)f(x) = \lim_{n\to\infty} \mathcal{R}(\Omega)f_n(x), \tag{7.2.2}$$

and call $\mathcal{R}(\Omega)f$ a *generalized solution* of the equation $Lu + f = 0$ in Ω with zero boundary condition.

EXERCISE* 7.2.5. Prove that $\mathcal{R}(\Omega)$ is a linear and bounded operator acting from $C(\bar{\Omega})$ into $C(\Omega)$.

In connection with this exercise it is interesting to observe that even if $L = \Delta$ and we define $\mathcal{R}(\Omega)1(x) = 0$ on $\partial\Omega$, the resulting function $\mathcal{R}(\Omega)1$ defined now in the whole $\bar{\Omega}$ generally speaking is *not* continuous. Also, if $L = \Delta$ and f is only continuous in $\bar{\Omega}$, even if $\Omega = B_1$, the function $\mathcal{R}(\Omega)f$ is not necessarily twice continuously differentiable in Ω (see Problem 4.9 in [**6**]).

EXAMPLE 7.2.6. Let $d \ge 3$, $L = \Delta$ and $\Omega = B_1 \setminus \{0\}$. Take $\Omega_n = B_1 \setminus \bar{B}_{1/n}$. One can easily check that $\mathcal{R}(\Omega_n)1(x)$, which are smooth solutions of $\Delta u + 1 = 0$ in Ω_n with zero boundary conditions, coincide with

$$u_n(x) = \gamma_n(1 - \frac{1}{|x|^{d-2}}) + \frac{1}{2d}(1 - |x|^2),$$

where the constant γ_n is chosen so that $u_n(x) = 0$ for $|x| = 1/n$. Obviously, $\gamma_n \to 0$ and $\mathcal{R}(\Omega_n)1(x) \to (1 - |x|^2)/(2d)$ in Ω. Therefore, to preserve the continuity of $\mathcal{R}(\Omega)1$ we should define $\mathcal{R}(\Omega)1$ at the boundary point $x = 0$ as $(2d)^{-1}$ rather than 0.

Much deeper examples of the same kind can be found in the literature.

EXERCISE 7.2.7. For $z \ge 0$ let $\mathcal{R}_z(\Omega)$ be the operator $\mathcal{R}(\Omega)$ corresponding to $L - z$. Assume that $\Omega \in C^{2+\delta}$ and prove that

$$\mathcal{R}_{z_1}(\Omega) = \mathcal{R}_{z_2}(\Omega) + (z_2 - z_1)\mathcal{R}_{z_1}(\Omega)\mathcal{R}_{z_2}(\Omega).$$

EXERCISE 7.2.8. Under the conditions and notation from Exercise 7.2.7 prove that if $f \in C(\bar\Omega)$, then $z|\mathcal{R}_z(\Omega)f|_{0;\Omega} \le |f|_{0;\Omega}$ for any $z > 0$; and if, in addition, $f = 0$ on $\partial\Omega$, then $|f - z\mathcal{R}_z(\Omega)f|_{0;\Omega} \to 0$ as $z \to \infty$.

EXERCISE 7.2.9 (also see Exercise 7.4.11). Under the conditions and notation from Exercise 7.2.7 prove that if $f \in C_0(\bar\Omega)$ and $\mathcal{R}(\Omega)f = 0$ in Ω, then $f = 0$ in Ω.

7.3. Generalized solutions of the Dirichlet problem with continuous boundary conditions

We take a bounded domain Ω and L as in Section 7.2. The operator $\mathcal{R}(\Omega)$ introduced in Definition 7.2.4 has the following properties.

LEMMA 7.3.1. *(i) If $f, g \in C(\bar\Omega)$ and $f \ge g$, then $\mathcal{R}(\Omega)f \ge \mathcal{R}(\Omega)g$. Moreover, $|\mathcal{R}(\Omega)f| \le \mathcal{R}(\Omega)|f|$.*

(ii) If $\Omega_1 \subset \Omega_2$ and $f, g \in C(\bar\Omega_2)$ and $f \ge g_+$, then $\mathcal{R}(\Omega_1)g \le \mathcal{R}(\Omega_2)f$ in Ω_1.

(iii) There exists a constant N depending only on κ, d and maxima of $|a^{ij}|, |b^i|$, $|c|$ and the diameter of Ω such that $|\mathcal{R}(\Omega)f|_{0;\Omega} \le N|f|_{0;\Omega}$.

(iv) Let Ω_n be a sequence of domains such that $\Omega_n \subset \Omega_{n+1} \subset \Omega$ and $\Omega = \cup_n\Omega_n$. Let $f \in C(\bar\Omega)$. Then for any $x \in \Omega$

$$\mathcal{R}(\Omega)f(x) = \lim_{n\to\infty} \mathcal{R}(\Omega_n)f(x). \tag{7.3.1}$$

Proof. Properties (i), (ii) and (iii) are trivial consequences of the definitions and discussion from Section 7.2.

In the proof of (iv) we first draw the reader's attention to the fact that the domains Ω_n are not required to lie strictly inside Ω. Also, observe that the equality $f = f_+ - f_-$ allows us to prove (7.3.1) only for $f \ge 0$, whereas assertion (iii) allows us to further reduce the general case to the one in which, in addition, f is smooth. In that case by (ii) the sequence $\mathcal{R}(\Omega_n)f(x)$ increases at any $x \in \Omega$ (starting with n such that $x \in \Omega_n$), and its limit is not greater than $\mathcal{R}(\Omega)f(x)$. On the other hand, take smooth domains Ω'_n such that $\Omega'_n \subset \bar\Omega'_n \subset \Omega$ and $\Omega = \cup_n\Omega'_n$. Then for any n there is m such that $\Omega'_n \subset \Omega_m$. Therefore, $\mathcal{R}(\Omega'_n)f \le \mathcal{R}(\Omega_m)f$ in Ω'_n, whence

$$\mathcal{R}(\Omega'_n)f \le \lim_{m\to\infty} \mathcal{R}(\Omega_m)f.$$

By letting $n \to \infty$ and applying Definition 7.2.2 we see that $\mathcal{R}(\Omega)f$ is not greater than the right-hand side of (7.3.1). The lemma is proved.

DEFINITION 7.3.2. If $g \in C^{2+\delta}(\bar\Omega)$ and $x \in \Omega$, then let

$$\pi(\Omega)g(x) = g(x) + \mathcal{R}(\Omega)(Lg)(x).$$

We will extend the operator $\pi(\Omega)$ to all continuous functions g given on $\bar\Omega$ by using the following result.

LEMMA 7.3.3. *(i) Let Ω_n be a sequence of domains such that $\Omega_n \subset \Omega_{n+1} \subset \Omega$ and $\Omega = \cup_n\Omega_n$. Let $g \in C^{2+\delta}(\bar\Omega)$. Then for any $x \in \Omega$*

$$\pi(\Omega)g(x) = \lim_{n\to\infty} \pi(\Omega_n)g(x).$$

(ii) If $g_1, g_2 \in C^{2+\delta}(\bar\Omega)$ and $g_1 \ge g_2$ in Ω, then $\pi(\Omega)g_1 \ge \pi(\Omega)g_2$ in Ω.

(iii) If $g \in C^{2+\delta}(\bar{\Omega})$ and $g \geq 0$, then $0 \leq \pi(\Omega)g \leq |g|_{0;\Omega}$ in Ω, and for any $g \in C^{2+\delta}(\bar{\Omega})$,

$$|\pi(\Omega)g|_{0;\Omega} \leq |g|_{0;\partial\Omega}. \tag{7.3.2}$$

Proof. The first statement follows from the definition of $\pi(\Omega)g$ and Lemma 7.3.1 (iv).

Next, we prove (ii) and (iii) in the case $\Omega \in C^{2+\delta}$. In this case by the definitions of $\mathcal{R}(\Omega)f$ and $\pi(\Omega)g$ we get that $\pi(\Omega)g \in C^{2+\delta}(\Omega)$ and $L\pi(\Omega)g = 0$ in Ω and $\pi(\Omega)g$ is continuously extended to $\partial\Omega$ if one defines $\pi(\Omega)g = g$ on $\partial\Omega$ (remember that initially $\pi(\Omega)g$ is defined only in Ω). From these properties of $\pi(\Omega)g$ assertions (ii) and (iii) follow by the maximum principle.

In the case of general bounded Ω assertion (ii) follows from the particular case and from (i). The same is true about assertion (iii), in which the first part is almost obvious. As for the second part, by taking smooth domains Ω_n in an obvious way, for any $x \in \Omega$ and n so large that $x \in \Omega_n$ and for appropriate $y_n \in \partial\Omega_n$ we get

$$|\pi(\Omega_n)g(x)| \leq \pi(\Omega_n)|g|(x) \leq \max_{\partial\Omega_n} |g|(y) = |g(y_n)|,$$

$$|\pi(\Omega)g|(x) = \lim_{n\to\infty} |\pi(\Omega_n)g(x)| \leq \limsup_{n\to\infty} |g(y_n)| \leq \max_{\partial\Omega} |g| = |g|_{0;\partial\Omega}.$$

This gives us (7.3.2). The lemma is proved.

Uryson's theorem and assertion (iii) of Lemma 7.3.3 show that the following definitions make sense.

DEFINITION 7.3.4. If g is a continuous function given on $\partial\Omega$, take any sequence of sufficiently smooth functions g_n defined on $\bar{\Omega}$ such that $|g - g_n|_{0;\partial\Omega} \to 0$ and for $x \in \Omega$ define

$$\pi(\Omega)g(x) = \lim_{n\to\infty} \pi(\Omega)g_n(x).$$

If f, g are continuous functions defined in $\bar{\Omega}$ and $\partial\Omega$ respectively, call the function

$$u(x) := \mathcal{R}(\Omega)f(x) + \pi(\Omega)g(x) \tag{7.3.3}$$

a *generalized solution* of the equation $Lu + f = 0$ in Ω with boundary data g.

EXERCISE* 7.3.5. Prove that the statements of Lemma 7.3.3 remain valid if we replace the space $C^{2+\delta}(\bar{\Omega})$ with $C(\bar{\Omega})$. Also prove that if $g_1, g_2 \in C(\bar{\Omega})$ and $g_1 \geq g_2$ on $\partial\Omega$, then $\pi(\Omega)g_1 \geq \pi(\Omega)g_2$ in Ω.

EXERCISE 7.3.6. Prove that if $c \equiv 0$, then $\pi(\Omega)1 \equiv 1$. (Actually $c \equiv 0 \iff \pi(\Omega)1 \equiv 1$.)

7.4. Some properties of generalized solutions

First of all we want to justify Definition 7.3.4.

DEFINITION 7.4.1. Let $a \in \partial\Omega$. We call the point a *regular* (relative to L and Ω) if

$$\lim_{\Omega\ni x\to a} \mathcal{R}(\Omega)1(x) = 0.$$

Notice that if $\Omega \in C^{2+\delta}$, then all points of $\partial\Omega$ are regular. Indeed, $\mathcal{R}(\Omega)1$ is a (unique) solution of class $C_0^{2+\delta}(\Omega)$ of the equation $Lu = -1$ in Ω.

THEOREM 7.4.2. *(i) Let $u \in C^2_{loc}(\Omega) \cap C(\bar{\Omega})$ and $f \in C(\bar{\Omega})$. Assume $f = -Lu$ in Ω. Then in Ω we have*

$$u = \mathcal{R}(\Omega)f + \pi(\Omega)u. \tag{7.4.1}$$

(ii) Let $g \in C(\partial\Omega), f \in C^\delta(\Omega)$. Then for the function u defined in (7.3.3) we have $u \in C^{2+\delta}_{loc}(\Omega)$. Furthermore, $Lu + f = 0$ in Ω and for any regular point $a \in \partial\Omega$ we have

$$\lim_{\Omega \ni x \to a} u(x) = g(a).$$

Proof. (i) For any domain $\Omega' \subset \bar{\Omega}' \subset \Omega$ we can find a sequence of functions $u_n \in C^{2+\delta}(\Omega')$ such that $|u - u_n|_{2;\Omega'} \to 0$ (for instance, one can mollify u). Then by definition of $\pi(\Omega')$

$$u_n = \mathcal{R}(\Omega')(-Lu_n) + \pi(\Omega')u_n$$

in Ω'. Upon letting $n \to \infty$ here and then $\Omega' \uparrow \Omega$ one easily gets (7.4.1) (also see Exercise 7.3.5).

(ii) First, let g be smooth. Then by definition $u = \mathcal{R}(\Omega)(f + Lg) + g$, which by Theorem 7.2.3 implies that $u \in C^{2+\delta}_{loc}(\Omega)$ and $Lu + f = 0$ in Ω. Theorem 7.1.1 on interior estimates yields that for any domain $\Omega' \subset \bar{\Omega}' \subset \Omega$ we have

$$|u|_{2+\delta;\Omega'} \le N(|f|_{\delta;\Omega} + |u|_{0;\Omega}) \le N(|f|_{\delta;\Omega} + |g|_{0;\Omega}), \tag{7.4.2}$$

where the constant N is independent of f, g. Also in that case

$$\lim_{\Omega \ni x \to a} |\mathcal{R}(\Omega)(f + Lg)(x)| \le N \lim_{\Omega \ni x \to a} \mathcal{R}(\Omega)1(x) = 0, \qquad \lim_{\Omega \ni x \to a} u(x) = g(a)$$

if a is regular.

If g is continuous only, we take an appropriate sequence of smooth g_n, define $u_n = \mathcal{R}(\Omega)f(x) + \pi(\Omega)g_n$, and notice that the assertions concerning the differential properties of u follow from (7.4.2). It only remains to notice that for any n

$$|u - u_n|_{0;\Omega} \le |g - g_n|_{0;\partial\Omega}, \quad \limsup_{\Omega \ni x \to a} |u(x) - g(a)|$$

$$\le \limsup_{\Omega \ni x \to a} |u - u_n|(x) + |g(a) - g_n(a)| \le 2|g - g_n|_{0;\partial\Omega}.$$

The theorem is proved.

DEFINITION 7.4.3. If the point a is regular and the function u is defined by (7.3.3), we set $u(a) = g(a)$.

COROLLARY 7.4.4. *If all the points of $\partial\Omega$ are regular and $g \in C(\partial\Omega), f \in C^\delta(\Omega)$, then the function u defined in (7.3.3) and in Definition 7.4.3 is a unique solution in $C^{2+\delta}_{loc}(\Omega) \cap C(\bar{\Omega})$ of the equation $Lu + f = 0$, which equals g on $\partial\Omega$.*

COROLLARY 7.4.5. *If f, g are continuous functions defined in $\bar{\Omega}$ and $\partial\Omega$ respectively and a domain $\Omega' \subset \bar{\Omega}' \subset \Omega$, then for the function u from (7.3.3) we have in Ω'*

$$u = \mathcal{R}(\Omega')f + \pi(\Omega')u. \tag{7.4.3}$$

Indeed, if f is smooth, then the function u is also sufficiently smooth and satisfies the equation $Lu + f = 0$ in Ω, which yields (7.4.3) by our definitions. In the general case, as usual, we make a passage to the limit.

REMARK 7.4.6. (This remark will not be used in the future.) In (7.4.3) we had to assume that $\bar{\Omega}' \subset \Omega$ in the first place, because u is not defined on all of $\partial\Omega$ but only at regular points (see Definition 7.4.3). However, if we define $u = g$ on $\partial\Omega$, then (7.4.3) becomes true for any $\Omega' \subset \Omega$ once one extends the operator $\pi(\Omega)$ to bounded and possibly discontinuous (Borel) functions given on Ω. Such an extension exists and is unique due to Riesz's representation theorem, by which for any fixed $x \in \Omega$ the linear bounded functional $\pi(\Omega)g(x)$ given on $C(\partial\Omega)$ can be written as

$$\pi(\Omega)g(x) = \int_{\partial\Omega} g(y)\pi_\Omega(x, dy),$$

where $\pi_\Omega(x, \cdot)$ is a measure on Borel subsets of $\partial\Omega$ (so–called L–harmonic measure). From the properties of $\pi(\Omega)$ it follows that $\pi_\Omega(x, \cdot)$ is nonnegative and $\pi_\Omega(x, \Omega) = \pi(\Omega)1(x) \le 1$.

The notion of regular points is local in the sense of the following theorem.

THEOREM 7.4.7. *Take a number $r > 0$. Then a point a is regular relative to L, Ω if and only if it is regular relative to $L, \Omega \cap B_r(a)$.*

Proof. If a is regular relative to L, Ω, then clearly it is regular relative to $L, \Omega \cap B_r(a)$, since $\mathcal{R}(\Omega)1 \ge \mathcal{R}(\Omega \cap B_r(a))1$ in $\Omega \cap B_r(a)$.

To prove the converse, let a be regular relative to $L, \Omega \cap B_r(a)$, and let Ω' be a subdomain of class $C^{2+\delta}$ of Ω. Observe, that (by Corollary 7.4.5)

$$\mathcal{R}(\Omega')1(x) = \mathcal{R}(\Omega' \cap B_r(a))1(x) + \pi(\Omega' \cap B_r(a))\mathcal{R}(\Omega')1(x)$$

in $\Omega' \cap B_r(a)$. Next, for a constant N, independent of Ω', we have $\mathcal{R}(\Omega')1(x) \le Nr$ if $x \in \Omega'$ (r is fixed), and $\mathcal{R}(\Omega')1 = 0$ on $\partial\Omega'$. It follows that $\mathcal{R}(\Omega')1(x) \le N|x - a| =: g(x)$ on $\partial(\Omega' \cap B_r(a))$. Therefore,

$$\mathcal{R}(\Omega')1(x) \le \mathcal{R}(\Omega' \cap B_r(a))1(x) + \pi(\Omega' \cap B_r(a))g(x)$$

in $\Omega' \cap B_r(a)$, and by letting $\Omega' \uparrow \Omega$ we get

$$\mathcal{R}(\Omega)1(x) \le \mathcal{R}(\Omega \cap B_r(a))1(x) + \pi(\Omega \cap B_r(a))g(x)$$

in $\Omega \cap B_r(a)$. It remains only to notice that if $\Omega \ni x \to a$, the right-hand side tends to zero by Theorem 7.4.2 (ii). The theorem is proved.

COROLLARY 7.4.8. *Let Ω_1, Ω_2 be bounded domains, $r > 0$, $a \in \partial\Omega_1$, and assume that $\Omega_1 \cap B_r(a) = \Omega_2 \cap B_r(a)$. Then a is regular relative to L, Ω_1 if and only if it is regular relative to L, Ω_2.*

COROLLARY 7.4.9 (exterior ball condition). *If there is a ball $B_\rho(x_0)$ such that $a = \bar{\Omega} \cap \bar{B}_\rho(x_0)$, then the point a is regular.*

Indeed, for simplicity let $x_0 = 0$ and observe that in $\Omega \cap B_{2\rho}$ we have

$$\mathcal{R}(\Omega \cap B_{2\rho})1(x) \leq \mathcal{R}(B_{2\rho} \setminus \bar{B}_\rho)1(x),$$

where the last expression is a $C_0^{2+\delta}(B_{2\rho} \setminus \bar{B}_\rho)$–solution of the equation $Lu + 1 = 0$. In particular, it tends to zero as x tends to any boundary point of $B_{2\rho} \setminus \bar{B}_\rho$, of which a is an example.

EXERCISE 7.4.10. Prove that if $\Omega \in C^{2+\delta}$, then at any point on $\partial\Omega$ an exterior ball condition is satisfied.

EXERCISE 7.4.11. Prove that if $f \in C(\bar{\Omega})$ and $\mathcal{R}(\Omega)f = 0$ in Ω, then $f = 0$ in Ω.

EXERCISE 7.4.12. On some occasions one prefers a different definition of regular points than Definition 7.4.1. Namely, one says that a point $a \in \partial\Omega$ is regular (for the operator L and domain Ω) if

$$\lim_{\Omega \ni x \to a} \pi(\Omega)g(x) = g(a)$$

for any bounded continuous g given on $\partial\Omega$. Show that this new definition is *equivalent* to Definition 7.4.1.

7.5. An example

Let $d = 2$ and $\Omega = (0, \pi)^2$, so that the exterior ball condition is satisfied at any point in $\partial\Omega$. Let us show that one can apply the above theory and explicitly find a unique $C_{loc}^{2+\delta}(\Omega) \cap C(\bar{\Omega})$–solution $u = u(x, y)$ of the equation $\Delta u + 1 = 0$ in Ω with zero boundary condition.

This solution is $u = \mathcal{R}(\Omega)1$. By "rounding up" the corners of Ω we can easily construct a sequence of smooth domains $\Omega_n \uparrow \Omega$ which differ from Ω only near the corners. Then $u_n := \mathcal{R}(\Omega_n)1 \to u$, and by Theorem 7.1.1 and Exercise 7.1.5 (i) for any n the norms $|u_p|_{2+\delta;\Omega_n}$ are uniformly bounded for $p \geq n$. It follows that $u \in C^{2+\delta}(\Omega_n)$ for any n and, say, $\nabla u_n \to \nabla u$ in Ω. Also the energy identity

$$\int_{\Omega_n} |\nabla u_n|^2 \, dxdy = -\int_{\Omega_n} u_n \Delta u_n \, dxdy = \int_{\Omega_n} u_n \, dxdy$$

shows that the first integral is bounded with respect to n. By Fatou's lemma this implies that

$$\infty > \int_\Omega |\nabla u|^2 \, dxdy = \int_0^{\pi/2} \Big(\int_{\partial\Omega(\varepsilon)} |\nabla u|^2 \, dS \Big) \, d\varepsilon,$$

where $\Omega(\varepsilon) = (\varepsilon, \pi - \varepsilon)^2$. Since the last integral converges, there is a sequence $\varepsilon_j \to 0$ such that

$$\varepsilon_j \int_{\partial\Omega(\varepsilon_j)} |\nabla u|^2 \, dS \to 0,$$

$$\int_{\partial\Omega(\varepsilon_j)} |\sin kx \, \sin my| \cdot |\nabla u| \, dS \leq (k + m)\varepsilon_j \int_{\partial\Omega(\varepsilon_j)} (1 + |\nabla u|^2) \, dS \to 0$$

for any integers k, m. We use this fact and remember that u is sufficiently smooth and vanishes on $\partial\Omega$ to carry out the following computations in which $\phi_{km} = \sin kx \sin my$:

$$0 = \int_\Omega \phi_{km}(1 + \Delta u)\,dxdy = (1 - (-1)^k)(1 - (-1)^m)\frac{1}{km} + I_{km},$$

where

$$I_{km} = \lim_{j\to\infty}\int_{\Omega(\varepsilon_j)} \phi_{km}\Delta u\,dxdy = -(k^2 + m^2)\lim_{j\to\infty}\int_{\Omega(\varepsilon_j)} u\phi_{km}\,dxdy$$

$$+ \lim_{j\to\infty}\int_{\partial\Omega(\varepsilon_j)} \left(\phi_{km}\frac{\partial}{\partial n}u - u\frac{\partial}{\partial n}\phi_{km}\right)dS = -(k^2 + m^2)\int_\Omega u\phi_{km}\,dxdy.$$

In this way we have found the Fourier coefficients of u. This function belongs to $L_2(\Omega)$. Therefore, almost everywhere

$$u(x, y) = \sum_{k,m=0}^{\infty} \frac{16}{(2k+1)(2m+1)}\,\frac{\sin(2k+1)x\,\sin(2m+1)y}{(2k+1)^2 + (2m+1)^2}. \tag{7.5.1}$$

Actually the series converges absolutely and uniformly, so that it is continuous, and hence (7.5.1) gives $u(x, y)$ for any $(x, y) \in \Omega$.

REMARK 7.5.1. Formula (7.5.1) can be used for computing values of u at various points. On the other hand, explicit formulas do not always give the best way to investigate properties of functions. For instance, from the maximum principle we know that $u \geq 0$. It appears nontrivial to get this very essential information on u from (7.5.1). Also we know that $u = \mathcal{R}1$ is infinitely differentiable in Ω, but formal differentiation of the right–hand side of (7.5.1) very soon leads to a series with growing coefficients.

EXERCISE 7.5.2. (i) By using (7.5.1) prove that $u \in C^1(\bar\Omega)$. (ii) Prove that $u \notin C^2(\bar\Omega)$.

EXERCISE 7.5.3. By using (7.5.1) prove that $u_x(0) = u_y(0) = 0$. In particular, u is not concave, as sometimes is expected for solutions of $\Delta u = -1$ vanishing on the boundary of a convex set.

7.6. Barriers and the exterior cone condition

The notion of regular point is related to the concept of *barrier* function.

DEFINITION 7.6.1. Let $a \in \partial\Omega$, $r > 0$ and define $\Omega_r = \Omega \cap B_r(a)$. Then a $C^2_{loc}(\Omega_r)$ function w is called a (local) *barrier* at a relative to L, Ω if:

$$Lw \leq 0 \quad \text{in} \quad \Omega \cap B_r(a), \qquad \inf_{x\in\Omega_r\setminus\Omega_\rho} w(x) > 0 \quad \forall\rho \in (0, r), \qquad \lim_{\Omega\ni x\to a} w(x) = 0.$$

EXERCISE 7.6.2. Prove that if a is regular, then there exists a (global) barrier at a.

EXERCISE 7.6.3. Let an exterior ball condition be satisfied at a point $a \in \partial\Omega$. Construct a barrier function *independent* of the operator L (but, of course, depending on d, K, κ).

THEOREM 7.6.4. *If there exists a barrier at a, then a is regular.*

Proof. Take r from Definition 7.6.1 and let a $C^{2+\delta}$–domain $\Omega' \subset \bar{\Omega}' \subset \Omega$. Also let $\rho \in (0,r)$ and $m_\rho := \inf\{w(x) : x \in \partial\Omega_\rho, |x-a| = \rho\}$. For a constant N_0 we have $\mathcal{R}(\Omega)1 \le N_0$ and $\mathcal{R}(\Omega')1 \le N_0 m_\rho^{-1} w$ on $\partial(\Omega' \cap B_\rho(a))$. Hence from the relations

$$w = -\mathcal{R}(\Omega' \cap B_\rho(a))Lw + \pi(\Omega' \cap B_\rho(a))w \ge \pi(\Omega' \cap B_\rho(a))w$$

we see that

$$\mathcal{R}(\Omega')1 = \mathcal{R}(\Omega' \cap B_\rho(a))1 + \pi(\Omega' \cap B_\rho(a))\mathcal{R}(\Omega')1$$

$$\le \mathcal{R}(\Omega' \cap B_\rho(a))1 + N_0 m_\rho^{-1}\pi(\Omega' \cap B_\rho(a))w \le \mathcal{R}(\Omega' \cap B_\rho(a))1 + N_0 m_\rho^{-1} w,$$

where the argument $x \in \Omega' \cap B_\rho(a)$ is dropped for simplicity. By letting $\Omega' \uparrow \Omega$ and then $x \to a$ we get

$$\limsup_{\Omega \ni x \to a} \mathcal{R}(\Omega)1(x) \le \limsup_{\Omega \ni x \to a} \mathcal{R}(\Omega \cap B_\rho(a))1(x). \tag{7.6.1}$$

Finally, we observe that $\mathcal{R}(\Omega \cap B_\rho(a))1 \le \mathcal{R}(B_\rho(a))1$, and by Exercise 6.1.4 we have $\mathcal{R}(B_\rho(a))1 \le N\rho^2$ if ρ is small, where N is independent of ρ. Therefore, the right–hand side of (7.6.1) tends to zero as $\rho \downarrow 0$. The theorem is proved.

In the proof of the following theorem we see one more extremely powerful idea.

THEOREM 7.6.5. *Let $a \in \partial\Omega$ and fix some numbers $r, \gamma > 0$. Assume that for any $\rho \in (0,r]$ the set $B_\rho(a) \setminus \Omega$ contains a ball of radius $\gamma\rho$. Then there are constants $N, \beta > 0$ such that $\mathcal{R}(\Omega)1(x) \le N|x-a|^\beta$ in Ω. In particular, the point a is regular.*

Proof. We may assume that $a = 0$. Define $m(\rho) = \sup\{\mathcal{R}(\Omega)1 : B_\rho \cap \Omega\}$. Also, fix ρ and denote x_ρ the center of the ball of radius $\gamma\rho$ lying in $B_\rho(a) \setminus \Omega$. It is clear that

$$B_\rho \subset B_{2\rho}(x_\rho) \subset B_{3\rho}(x_\rho) \subset B_{4\rho}.$$

Now, we take a smooth domain $\Omega' \subset \bar{\Omega}' \subset \Omega$ such that $B_\rho \cap \Omega' \ne \emptyset$ and a point $x \in B_\rho \cap \Omega'$. We have

$$\mathcal{R}(\Omega')1(x) = \mathcal{R}(B_{3\rho}(x_\rho) \cap \Omega')1(x) + \pi(B_{3\rho}(x_\rho) \cap \Omega')\mathcal{R}(\Omega')1(x). \tag{7.6.2}$$

The first term on the right is less than $\mathcal{R}(B_{3\rho}(x_\rho))1(x)$, which is less than $N\rho^2$ by Exercise 6.1.4. In the second term we have $\mathcal{R}(\Omega')1 = 0$ on $B_{3\rho}(x_\rho) \cap \partial\Omega'$ and

$$\mathcal{R}(\Omega')1 \le \sup_{B_{3\rho}(x_\rho) \cap \Omega'} \mathcal{R}(\Omega')1 \le \sup_{B_{4\rho} \cap \Omega'} \mathcal{R}(\Omega')1 \le m(4\rho)$$

on $(\partial B_{3\rho}(x_\rho)) \cap \Omega'$. Furthermore, define $u(x) = \pi(B_{3\rho}(x_\rho) \setminus \bar{B}_{\gamma\rho}(x_\rho))g(x)$, where $g = 1$ on $\partial B_{3\rho}(x_\rho)$ and $g = 0$ on $\partial B_{\gamma\rho}(x_\rho)$. Then u is a unique smooth solution of the equation $Lu = 0$ in $B_{3\rho}(x_\rho) \setminus \bar{B}_{\gamma\rho}(x_\rho)$, which takes the boundary data g. Obviously, $\mathcal{R}(\Omega')1 \le m(4\rho)u$ on $\partial(B_{3\rho}(x_\rho) \cap \Omega')$. From (7.6.2) it follows that

$$\mathcal{R}(\Omega')1(x) \le N\rho^2 + m(4\rho)\pi(B_{3\rho}(x_\rho) \cap \Omega')u(x) = N\rho^2 + m(4\rho)u(x).$$

By letting $\Omega' \uparrow \Omega$ we get

$$m(\rho) \le N\rho^2 + m(4\rho) \sup_{B_{2\rho}(x_\rho)} u(x),$$

where the last supremum is less than a constant $\varepsilon < 1$ by Exercise 6.1.5.

Thus for $\rho \in (0,1]$ we have

$$m(\rho) \le N_1\rho^2 + \varepsilon m(4\rho),$$

where the constants N_1, ε depend only on d, K, κ. We certainly may assume that $\varepsilon \ge 1/8$. In that case for the function $M(\rho) := m(\rho) + N_1\rho^2$ we have $M(\rho) \le \varepsilon M(4\rho)$, $M(\rho/4) \le \varepsilon M(\rho)$, $M(\rho/4^n) \le \varepsilon^n M(\rho)$, $M(4^{-n}) \le \varepsilon^n N$,

$$m(4^{-n}) \le \varepsilon^n N, \quad \forall n = 0, 1, 2, \dots.$$

Finally, we observe that m is a nondecreasing function and $\rho \le 4^{-n}$ if $n = [\log_4(1/\rho)]$. Therefore, for $\rho \in (0,1]$

$$m(\rho) \le N\varepsilon^n \le N\varepsilon^{\log_4(1/\rho)-1} = N\rho^\beta,$$

where $\beta = -\log_4 \varepsilon \in (0,1)$. The theorem is proved.

REMARK 7.6.6. The condition of this theorem is satisfied, for instance, if the so–called *exterior cone condition* is satisfied: there exists a finite right circular cone $\mathcal{K}$, with vertex a, satisfying $\bar{\mathcal{K}} \cap \bar{\Omega} = a$.

EXERCISE 7.6.7. For $r > 0$ define $u_r(x) = \pi(B_r(a) \cap \Omega)\phi_r(x)$, where $\phi_r(x) = |x-a|/r$. Prove that

(i) If the point a is regular, then

$$\lim_{r\downarrow 0} \lim_{\Omega \ni x \to a} u_r(x) = 0.$$

(ii) In $B_r(a) \cap \Omega$ we have $u_r \le 1$, and if

$$\xi := \limsup_{r\downarrow 0} \limsup_{\Omega \ni x \to a} u_r(x) < 1,$$

then a is regular, so that $\xi = 0$.

In other words prove the following "zero-one law" and De la Vallée Poussin's criterion: We always have $\xi = 0$ or $\xi = 1$, and the point a is regular if and only if $\xi = 0$.

REMARK 7.6.8. De la Valle Poussin's criterion is known to be useful to prove that regular points for the operators L with coefficients satisfying so-called Dini's continuity condition coincide with regular points for the Laplacian. For the Laplacian there is a celebrated Wiener's criterion.

REMARK 7.6.9. Given a domain Ω, the set of regular points *depends* on the operator under consideration. For example, take $d \ge 2$, the domain Ω from Example 7.2.6 and the operator

$$L = \Delta + \mu \sum_{i,j=1}^{d} \frac{x^i x^j}{|x|^2} \frac{\partial^2}{\partial x^i x^j},$$

where the constant $\mu > d-2$. The coefficients of L are infinitely differentiable outside zero, and one can easily define $\mathcal{R}(\Omega)$ and the notion of regular point. But in contrast with Example 7.2.6 the origin is *regular* with respect to L.

Indeed, for $\Omega_n = B_1 \setminus \bar{B}_{1/n}$ the function $\mathcal{R}(\Omega_n)1$ is a unique smooth solution of the equation $Lu + 1 = 0$ in Ω_n with zero boundary condition. Looking for spherically symmetric solutions, one readily gets

$$\mathcal{R}(\Omega_n)1(x) = \tau_n(1-|x|^\gamma) + \frac{1}{2(d+\mu)}(1-|x|^2),$$

$$\gamma = \frac{\mu+2-d}{\mu+1}, \qquad \tau_n = \frac{1-n^{-2}}{2(d+\mu)(n^{-\gamma}-1)},$$

$$\mathcal{R}(\Omega)1(x) = -\frac{1}{2(d+\mu)}(1-|x|^\gamma) + \frac{1}{2(d+\mu)}(1-|x|^2),$$

which tends to zero as $x \to 0$.

7.7. Hints to exercises

7.1.5. (i) Repeat the proof of Theorem 7.1.1 by replacing $|u|_{k+m+\delta;B_{R_n}}$ with $|u\zeta_n|_{k+m+\delta;\Omega}$. Do not miss that this time the norms $|u\zeta_n|_{k+m+\delta;\Omega}$ go to infinity as $n \to \infty$. You can find more detail in the proof of Theorem 8.11.1.

(ii) One can follow the proof of Theorem 7.1.2, shrinking B_{2R} to B_R in k steps, applying solvability results for Ω, and taking cut–off functions ζ such that $\zeta = 1$ in B_R and $\zeta = 0$ outside B_{2R}.

7.1.6. Solve the equation $Lu + f = 0$ in B_R with zero boundary conditions and send $R \to \infty$. Also apply the result of Exercise 6.1.3.

7.2.5. For $f \in C^\delta(\Omega)$ we have the continuity of $\mathcal{R}(\Omega)f$ in Ω from Theorem 7.2.3. For general $f \in C(\bar{\Omega})$ prove that the convergence in (7.2.2) is uniform in Ω.

7.2.7. See the hint to Exercise 4.4.5.

7.2.8. Reduce the situation to the case of smooth f. Then notice that by definition $z\mathcal{R}_z(\Omega)f = \mathcal{R}_z(\Omega)Lf + f$ if $f \in C_0^{2+\delta}(\Omega)$.

7.2.9. Apply Exercises 7.2.7 and 7.2.8.

7.4.11. Apply Corollary 7.4.5 to small balls in $\Omega' = \{f > 0\} \cap \Omega$ assuming that this set is not empty. Also notice that as follows from the maximum principle, for small $r > 0$ we have $\mathcal{R}(B_r)1(x) \geq c(r^2 - |x|^2)$ where the constant $c > 0$ is independent of x, r.

7.4.12. In one direction use Theorem 7.4.2 (ii). To prove the converse use Theorem 7.4.2 (i) for the function v_0 from Lemma 6.1.1.

7.5.2. (ii) What should be the value of Δu at the origin if $u \in C^2(\bar{\Omega})$? Remember that $u = 0$ on $\partial\Omega$.

7.6.2. Consider $\varepsilon|x-a|^2 + \mathcal{R}(\Omega)1(x)$.

7.6.3. See the hint to Exercise 6.1.5.

7.6.7. To prove (ii) first prove that if $x \in B_r(a) \cap \Omega$, then

$$\mathcal{R}(\Omega)1(x) \leq \mathcal{R}(B_r(a)\cap\Omega)1(x) + u_r(x) \sup_{B_r(a)\cap\Omega} \mathcal{R}(\Omega)1.$$

Then conclude that $\limsup_{\Omega\ni x\to a} \mathcal{R}(\Omega)1(x) \leq \xi \limsup_{\Omega\ni x\to a} \mathcal{R}(\Omega)1(x)$.

CHAPTER 8

Parabolic Equations in the Whole Space

In this chapter, we start considering parabolic second–order equations of the type

$$\frac{\partial}{\partial t}u(t,x) = L(t,x)u(t,x) + f(t,x), \tag{8.0.1}$$

where

$$L(t,x) = a^{ij}(t,x)\frac{\partial^2}{\partial x^i \partial x^j} + b^i(t,x)\frac{\partial}{\partial x^i} + c(t,x) \tag{8.0.2}$$

is a second–order elliptic (with respect to x) operator given for any $t \in (-\infty, \infty)$ and $x \in \mathbb{R}^d$. In particular, we assume that a, b, c are real valued and $c \leq 0$. The latter assumption is not a restriction at all, since for any constant λ the function $v(t,x) = u(t,x)e^{-\lambda t}$ satisfies $v_t = Lv - \lambda v + fe^{-\lambda t}$ if u satisfies (8.0.1).

The *heat equation*

$$\frac{\partial}{\partial t}u(t,x) = \Delta u(t,x) + f(t,x)$$

is a particular case of equations to be considered. Notice that by Δ we always mean Laplace's operator applied with respect to the variable x.

The reader who did not follow the parts of the book where we discuss equations with complex coefficients can skip Secs. 8.2 and 8.3. The rest of the book is independent of these two sections.

8.1. The maximum principle

Let Ω be a domain in $\mathbb{R}^d$.

DEFINITION 8.1.1. Take $-\infty \leq S < T \leq \infty$ and denote $Q = (S,T) \times \Omega$. Define the *parabolic boundary* $\partial' Q$ of Q as the set

$$[(S,T) \times \partial\Omega] \cup \{(t,x) \in \mathbb{R}^{d+1} : t = S, x \in \bar{\Omega}\}.$$

Also denote

$$\partial_t Q = \{(t,x) \in \mathbb{R}^{d+1} : t = S, x \in \bar{\Omega}\}, \quad \partial_x Q = (S,T) \times \partial\Omega,$$

$$\partial_{tx} Q = \{(t,x) \in \mathbb{R}^{d+1} : t = S, x \in \partial\Omega\}.$$

This notation is natural if one interprets ∂ as a limit of corresponding increments.

THEOREM 8.1.2. *(i) Let $T \in (0,\infty), Q = (0,T) \times \Omega$. (ii) Assume that in $\bar{Q}$ we are given a function u which is bounded and continuous and for any t the derivatives $u_x(t,x), u_{xx}(t,x)$ exist and are continuous in Ω and the derivative u_t exists at any point in Q. (iii) Let Ω be a bounded domain. (iv) Also assume that $Lu - u_t \geq 0$ in Q and $u \leq 0$ on $\partial' Q$. Then $u \leq 0$ in Q.*

Proof. Take a constant $\gamma > 0$ and define $v = u - \gamma/(T-t)$. Also let z_γ be a point in $\bar{Q}$ at which v takes its maximum value. Observe that $v(z) \to -\infty$ as z approaches the upper lid of Q. Therefore, $z_\gamma \in Q \cup \partial' Q$. Actually if $v(z_\gamma) \geq 0$, then z_γ cannot lie inside Q. Indeed, if it does, then the matrix $v_{xx}(z_\gamma) \leq 0$ and $v_t(z_\gamma) = v_x(z_\gamma) = 0$, so that at z_γ we have

$$0 \leq Lu - u_t = Lu - \gamma/(T-t)^2 \leq -\gamma/(T-t)^2 + cu$$
$$\leq -\gamma/(T-t)^2 + c\gamma/(T-t) < 0,$$

which is impossible.

Thus either $v(z_\gamma) < 0$ or $z_\gamma \in \partial' Q$, in which case $v(z_\gamma) \leq 0$. We see that always $v(z_\gamma) \leq 0$ and $v \leq 0$ in Q. Since γ is arbitrary, the theorem is proved.

COROLLARY 8.1.3. *Under the assumptions (i), (ii) and (iii) of Theorem 8.1.2 we have*

$$|u|_{0;Q} \leq T|Lu - u_t|_{0;Q} + |u|_{0;\partial' Q}. \tag{8.1.1}$$

In particular, (uniqueness) if $u = 0$ *on* $\partial' Q$ *and* $Lu - u_t = 0$ *in* Q, *then* $u \equiv 0$ *in* Q.

Indeed, we may assume that the right–hand side of (8.1.1) is finite and denote $N_0 = |Lu - u_t|_{0;Q}$, $N_1 = |u|_{0;\partial' Q}$. Then for the function $v = u - N_1 - N_0 t$ we have $Lv - v_t \geq 0$ in Q and $v \leq 0$ on $\partial' Q$. Therefore $u \leq N_1 T + N_0$ in Q. Similarly, $-u \leq N_1 T + N_0$.

Next we extend Theorem 8.1.2 to unbounded domains Ω.

THEOREM 8.1.4. *(i) Let* $T \in (0, \infty), Q = (0,T) \times \Omega$. *(ii) Assume that in* $\bar{Q}$ *we are given a function* u *which is bounded and continuous and for any* t *the derivatives* $u_x(t,x), u_{xx}(t,x)$ *exist and are continuous in* Ω *and the derivative* u_t *exists at any point in* Q. *(iii) Let the coefficients of* L *be bounded in* Q. *(iv) Also assume that* $Lu - u_t \geq 0$ *in* Q *and* $u \leq 0$ *on* $\partial' Q$. *Then* $u \leq 0$ *in* Q.

Proof. It is easy to check that for a constant $\lambda > 0$ large enough and for the function $v_0 = (\cosh|x|)\exp(\lambda t)$ we have $Lv_0 - v_{0t} \leq 0$. Define $m = |u|_{0;Q}$, $Q_{T,R} := (0,T) \times [\Omega \cap B_R]$. Then the function $w_R = u - v_0 m(\cosh R)^{-1}$ satisfies $w_R \leq 0$ on $\partial' Q_{T,R}$, $Lw_R - w_{Rt} \geq 0$. By Theorem 8.1.2 we have $u \leq v_0 m(\cosh R)^{-1}$ in $Q_{T,R}$. By letting $R \to \infty$ we get $u \leq 0$, and the theorem is proved.

Similarly to Corollary 8.1.3 we get the following.

COROLLARY 8.1.5. *Under the assumptions (i), (ii) and (iii) of Theorem 8.1.4 we have*

$$|u|_{0;Q} \leq T|Lu - u_t|_{0;Q} + |u|_{0;\partial' Q}.$$

In particular, (uniqueness) if $u = 0$ *on* $\partial' Q$ *and* $Lu - u_t = 0$ *in* Q, *then* $u \equiv 0$ *in* Q.

The following corollary says that solutions of parabolic equations up to time $S \leq T$ depend only on values of data given up to S too.

COROLLARY 8.1.6. *Take* $S \in (0,T)$ *and denote* $Q' = (0,S) \times \Omega$. *Let the assumptions (i), (ii) and (iii) of Theorem 8.1.2 or Theorem 8.1.4 be satisfied. Also assume that in* $\bar{Q}'$ *we are given a function* v *which is bounded and continuous in* $\bar{Q}'$ *and for any* $t \in (0,S)$ *the derivatives* $v_x(t,x), v_{xx}(t,x)$ *exist and are continuous in* Ω *and the derivative* v_t *exists at any point in* Q'. *Finally, let* $v = u$ *on* $\partial' Q'$ *and* $Lu - u_t = Lv - v_t$ *in* Q'. *Then* $u = v$ *in* $\bar{Q}'$.

The next theorem gives an estimate of $|u|_{0;Q}$ independent of T.

THEOREM 8.1.7. *Let $T \in (0, \infty], Q = (0, T) \times \Omega$. Then under the assumptions (ii) and (iii) of Theorem 8.1.2 or Theorem 8.1.4, for any $\lambda > 0$ we have*

$$|u|_{0;Q} \leq \lambda^{-1}|Lu - \lambda u - u_t|_{0;Q} + |u|_{0;\partial' Q}. \tag{8.1.2}$$

The same estimate holds if $T \in (-\infty, \infty]$ and $Q = (-\infty, T)\times\Omega$ and the assumptions (ii) and (iii) of Theorem 8.1.4 are satisfied (we set $|u|_{0;\partial' Q} = 0$ if $\Omega = \mathbb{R}^d$).

Proof. Owing to an obvious possibility to pass to the limit, we may assume that $T < \infty$. Also we may assume that the right–hand side of (8.1.2) is finite. To prove the first assertion, define

$$M = |Lu - \lambda u - u_t|_{0;Q}, \quad m = |u|_{0;\partial' Q}, \quad v = e^{\lambda t}[u - m - \lambda^{-1}M].$$

Then obviously $v \leq 0$ on $\partial' Q$ and

$$\begin{aligned} Lv - v_t = e^{\lambda t}[Lu - \lambda u - u_t] - ce^{\lambda t}[m + \lambda^{-1}M] + e^{\lambda t}[\lambda m + M] \\ \geq e^{\lambda t}[Lu - \lambda u - u_t + M] \geq 0. \end{aligned}$$

Hence, $v \leq 0$ and $u \leq m + \lambda^{-1}M$ in Q. Similarly $-u \leq m + \lambda^{-1}M$ in Q.

To prove the second assertion, apply the first one to $v = e^{\varepsilon t}u$, where $\varepsilon > 0$, in the domain $Q_{S,T} = (S, T) \times \Omega$ instead of $(0, T) \times \Omega$. Then

$$|v|_{0;Q_{S,T}} \leq \lambda^{-1}|e^{\varepsilon t}[Lu - (\lambda + \varepsilon)u - u_t]|_{0;Q} + e^{\varepsilon T}|u|_{0;\partial' Q} + e^{\varepsilon S}\sup_{\bar\Omega}|u(S, x)|.$$

By letting $S \to -\infty$ we get that in Q

$$e^{\varepsilon t}|u(t, x)| \leq \lambda^{-1}e^{\varepsilon T}|Lu - \lambda u - u_t|_{0;Q} + \lambda^{-1}e^{\varepsilon T}\varepsilon|u|_{0;Q} + e^{\varepsilon T}|u|_{0;\partial' Q}.$$

We now let $\varepsilon \downarrow 0$. The theorem is proved.

In the future we will need one more version of the above results.

THEOREM 8.1.8. *Let $T \in (-\infty, \infty]$, Ω be a bounded domain and $C := (-\infty, T) \times \Omega$. Assume that in $\bar C$ we are given a function u which is bounded and continuous and that for any $t \in (-\infty, T)$ the derivatives $u_x(t, x), u_{xx}(t, x)$ exist and are continuous in Ω and the derivative u_t exists at any point in C. Also assume that for certain constants $K, \kappa \in (0, \infty)$ we have*

$$\operatorname{tr} a(z) + |b(z)| + |c(z)| \leq K, \quad a^{ij}(z)\xi^i\xi^j \geq \kappa|\xi|^2$$

for any $z \in \mathbb{R}^{d+1}$ and $\xi \in \mathbb{R}^d$ (and as always $c(z) \leq 0$). Then

$$|u_\pm|_{0;C} \leq N|(u_t - Lu)_\pm|_{0;C} + |u_\pm|_{0;\partial' C}, \quad |u|_{0;C} \leq N|u_t - Lu|_{0;C} + |u|_{0;\partial' C},$$

where $N = N(d, K, \kappa, d_\Omega)$ and, of course, $\partial' C = (-\infty, T) \times \partial\Omega$.

Proof. Obviously we need to prove only the first inequality and only for sign $+$. Also we may assume that $T < \infty$ and $0 \in \Omega$. Then take the function v_0 from Lemma 6.1.1 corresponding to R such that $\Omega \subset B_{R/2}$. Also let $M = |(u_t - Lu)_+|_{0;C}$, $m = \sup_\Omega v_0$. We have $Lv_0 \leq -1$ in C and $v_0 \geq \gamma$ in C, where the constant $\gamma > 0$ depends only on d, κ, K, R. The function $w = v_0 \exp(-t/m)$ satisfies $Lw - w_t \leq 0$ in C. Next, for large $n > 0$ define $C_n = (-n, T) \times \Omega$

$$\eta_n := \sup_\Omega \frac{u_+}{w}(-n, x) \leq e^{-n/m}\gamma^{-1} \sup_C |u|.$$

For the function $\phi := u - Mv_0 - \eta_n w - |u_+|_{0;\partial' C}$ we have

$$\phi_t - L\phi = u_t - Lu + MLv_0 + \eta_n(Lw - w_t) + c|u_+|_{0;\partial' C} \leq M + MLv_0 \leq 0$$

in C_n and $\phi \leq 0$ on $\partial' C_n$. By Theorem 8.1.2 applied to the cylinder C_n we have $\phi \leq 0$ in C_n, so that

$$u(t, x) \leq Mm + \eta_n w(t, x) + |u_+|_{0;\partial' C}$$

whenever $t \geq -n$, $x \in \Omega$. It remains only to let $n \to \infty$. The theorem is proved.

REMARK 8.1.9. If the coefficients of L are growing, the assertions of Theorem 8.1.4 and Corollary 8.1.5 may be false for $\Omega = \mathbb{R}^d$. It turns out that for $d = 1$ the Cauchy problem

$$u'' + x^3 u' - u - u_t = 0 \quad \text{on} \quad (0, \infty) \times (-\infty, \infty), \quad u(0, x) \equiv 0$$

has a nontrivial bounded solution which is continuous in $[0, \infty) \times (-\infty, \infty)$. Also the same is true if one replaces x^3 by $|x|^r \operatorname{sign} x$ with any $r > 1$.

EXERCISE 8.1.10. Under the assumptions (i), (ii) and (iii) of Theorem 8.1.2 or Theorem 8.1.4 define $f = Lu - u_t$. Assume that $f \leq 0$ in Q and $u = 0$ on $\partial' Q$. Finally, let the coefficients of L and f be *independent* of t. Prove that $u_t \geq 0$ in Q.

EXERCISE 8.1.11. Suppose $u(x, t)$ satisfies

$$0 \leq u \leq M, \quad u_t = \Delta u - u^p \quad \text{in } \mathbb{R}^d \times (0, \infty)$$

with constants $M > 0$, $0 < p < 1$. By using Theorem 8.1.4 and comparing u with a solution of $v_t + v^p = 0$ show that $u \equiv 0$ on $\mathbb{R}^d \times (T, \infty)$ for some $T = T(M, p) > 0$.

EXERCISE 8.1.12 (computations for Exercise 8.1.13). Assume that we are given C^2–functions $\psi^1 = \psi^i(x)$, $i = 1, ..., n$, such that $\Omega = \{x : \psi_i(x) > 0, i = 1, ..., n\}$ is a nonempty bounded domain and $|\operatorname{grad} \psi_i| \geq 1$ on $\Gamma_i = \{\psi_i = 0\} \cap \partial\Omega$. Let $a^{ij}(t, x)\xi^i\xi^j \geq \kappa|\xi|^2$ for any t, x, ξ, with a constant $\kappa > 0$, and also assume that the coefficients of L are bounded. By a direct computation prove that for sufficiently large constant $\lambda > 0$ and sufficiently small constant $\varepsilon > 0$ the functions $\phi_i(t, x) = t^{-\varepsilon\kappa} \exp(\lambda t - \varepsilon\psi_i^2(x)/t)$ for $t > 0$ and $x \in \Omega$ satisfy $L\phi_i - \phi_{it} \leq 0$.

EXERCISE 8.1.13 (uniqueness of discontinuous solutions). Let the conditions of Exercise 8.1.12 be satisfied. Take $T \in (0, \infty)$ and denote $Q = (0, T) \times \Omega$ and assume that in $\bar{Q} \setminus \partial_{tx} Q$ we are given a function u which is bounded and continuous and that for any $t \in (0, T)$ the derivatives $u_x(t, x)$, $u_{xx}(t, x)$ exist and are continuous in Ω and the derivative u_t exists at any point in Q. Also assume that $Lu = u_t$ in Q and $u(0, x) = 0$ for $x \in (\partial' Q) \setminus \partial_{tx} Q$. Prove then that $u \equiv 0$ in Q.

The following three exercises are similar to Exercises 2.6.7, 2.6.9 and 2.6.10.

EXERCISE 8.1.14. Take $d = 1$ and find a function $u \not\equiv 0$ such that $u_t = u_{xx}$ in $\mathbb{R} \times (-\pi, \pi)$ and $u(t, \pm\pi) \equiv 0$. The existence of such functions shows that the assumption on boundedness of u in Theorem 8.1.8 is essential.

EXERCISE 8.1.15. Let $T \in (-\infty, \infty]$, $Q = (-\infty, T) \times \{-\pi/4 < x^d < \pi/4\}$ and u be a bounded continuous function in $\bar{Q}$ having continuous derivatives u_t, u_x, u_{xx}. Assume that $u_t = \Delta u$ in Q. Prove that

$$\sup_Q |u| = \sup_{\partial' Q} |u|. \tag{8.1.3}$$

EXERCISE 8.1.16. In Exercise 8.1.15 take $Q = (-\infty, T) \times \mathbb{R}^d_+$ and keep all other assumptions. Prove that (8.1.3) holds again.

In solving the following two exercises we suggest using the technique based on barriers. This technique can be carried over the case of nonlinear equations.

EXERCISE 8.1.17. Let $\delta \in (0, 1)$, $T \in (-\infty, \infty]$ and $Q = (-\infty, T) \times \mathbb{R}^d_+$. Let u satisfy the assumption of Theorem 8.1.8 and let $\Delta u = u_t$ in Q. Prove that

(i) if $|u(t, x) - u(s, x)| \leq |t - s|^\delta$ whenever $t, s \leq T$ and $x^d = 0$, then $|u(t, x) - u(s, x)| \leq |t - s|^\delta$ for all $t, s \leq T$ and $x \in \mathbb{R}^d_+$;

(ii) if $d \geq 2$ and $|u(t, x) - u(t, y)| \leq |x - y|^\delta$ whenever $t \leq T$ and $x^d = y^d = 0$, then $|u(t, x) - u(t, y)| \leq |x - y|^\delta$ for all $t \leq T$ and $x, y \in \mathbb{R}^d_+$ such that $x^d = y^d$;

(iii) if $d \geq 2$ and $|u(t, x) - u(s, x)| \leq |t - s|^{\delta/2}$ whenever $t, s \leq T$ and $x^d = 0$ and if $|u(t, x) - u(t, y)| \leq |x - y|^\delta$ whenever $t \leq T$ and $x^d = y^d = 0$, then $|u(t, x) - u(s, y)| \leq N(|x - y|^\delta + |t - s|^{\delta/2})$ for all $t, s \leq T, x, y \in \mathbb{R}^d_+$, where $N = N(d, \delta)$.

EXERCISE 8.1.18. Let $\delta \in (0, 1)$, $T \in (0, \infty]$ and $Q = (0, T) \times \mathbb{R}^d$. Let u satisfy assumption (iii) of Theorem 8.1.4, and let $\Delta u = u_t$ in Q. Prove that if $|u(0, x) - u(0, y)| \leq |x - y|^\delta$ for all $x, y \in \mathbb{R}^d$, then $|u(t, x) - u(t, y)| \leq |x - y|^\delta$ for all $x, y \in \mathbb{R}^d$ and also $|u(t, x) - u(s, x)| \leq N(d)|t - s|^{\delta/2}$ whenever $t, s \geq 0$ and $x \in \mathbb{R}^d$.

EXERCISE 8.1.19 (cf. Exercise 4.3.6). Let $F(\xi)$ be a continuously differentiable real-valued function on $\mathbb{R}^d$ such that $F(0) = 0$. Prove that for any bounded function $u = u(t, x)$ having bounded and continuous derivatives u_x, u_{xx}, u_t in $\mathbb{R}^{d+1}$ we have $|u|_0 \leq |f|_0$, where $f := \Delta u + F(\operatorname{grad} u) - u - u_t$. Also prove that for any function g any two bounded functions having bounded and continuous derivatives with respect to x, xx, t and satisfying the equation $\Delta u + F(\operatorname{grad} u) - u - u_t = g$ in $\mathbb{R}^{d+1}$ coincide.

EXERCISE 8.1.20 (a particular case of Chebyshev's inequality). Under the assumptions (i), (ii) and (iii) of Theorem 8.1.2 or Theorem 8.1.4 define $f = Lu - u_t$. Assume that $f \geq 0$ in Q and $u \leq 0$ on $\partial_x Q$ and $u \leq 1$ on $\partial_t Q$. Also let $v = v(x)$ be a smooth nonnegative in $\bar{\Omega}$ function such that $Lv \leq -1$ in Q. Prove that $u(t, x) \leq v(x)/t$ in Q.

EXERCISE 8.1.21. In the situation of Exercise 8.1.20 show that there exist finite constants $N, \lambda > 0$ such that $u(t, x) \leq Ne^{-\lambda t}$.

EXERCISE 8.1.22. Prove that the assertions of Theorem 8.1.4 and Corollary 8.1.5 remain valid if instead of boundedness of the coefficients of L we assume that $\operatorname{tr} a(t, x) + x \cdot b(t, x) \leq K(1 + |x|^2)$ in Q, where K is a constant.

EXERCISE 8.1.23 (Tychonoff's theorem). Let u be a continuous function in $[0, T] \times \mathbb{R}^d$ having continuous derivatives u_t, u_x, u_{xx} in $(0, T) \times \mathbb{R}^d$. Let $|u| \leq N \exp(N|x|^2)$ in $[0, T] \times \mathbb{R}^d$, where N is a constant, and $u_t = \Delta u$ in $(0, T) \times \mathbb{R}^d$ and $u(0, x) \equiv 0$. Prove that $u \equiv 0$.

REMARK 8.1.24. Widder's theorem says that any *nonnegative* function which is continuous in $[0, \infty) \times \mathbb{R}$ and equals zero for $t = 0$ and satisfies the heat equation $u_{xx} - u_t = 0$ in $(0, \infty) \times \mathbb{R}$ is identically zero.

Tychonoff's example shows that the assumption on sign of u is essential. Namely, the following function satisfies all the above requirements apart from positivity:

$$u(t, x) = \sum_{k=0}^{\infty} \frac{g^{(k)}(t)}{(2k)!} x^{2k}, \quad g(t) = \exp(-t^{-2}) \quad t > 0, \quad g(0) = 0.$$

8.2. The Cauchy problem, semigroup approach, motivation

In this section as always we fix a $\delta \in (0, 1)$. We also take a constant $K > 0$ and a second–order elliptic operator L with coefficients a, b, c $(c \leq 0)$ independent of t and the ellipticity constant $\kappa > 0$ and assume that $|a, b, c|_\delta \leq K$. In particular, we assume that $a, b, c \in C^\delta(\mathbb{R}^d)$. Our goal is to solve the Cauchy problem:

$$\frac{\partial u}{\partial t} = Lu \quad t > 0, \quad u(0, \cdot) = f. \tag{8.2.1}$$

First of all we want to explain why what follows in this section is natural. We want to look at (8.2.1) as an equation in an appropriate Banach space treating functions as vectors. Recall that for finite–dimensional vector–valued functions u and square matrix L the following very inspiring formula turned out to be useful:

$$u(t) = e^{tL} f.$$

We want to use the same formula. In finite–dimensional space it is possible to define $\exp(tL)$ by using Taylor's series. In our case, however, the expression $L^n f$ makes little sense, since the coefficients of L may only be Hölder continuous and in the expression $L^2 f = L(Lf)$ their derivatives are involved (however, in connection with this see Remark 8.3.2). In any case the series for $e^{tL} f$ resembles Taylor's series for f, and its convergence seems to be related to a very high regularity of f.

Therefore, we need a different formula. Remember that again in the matrix case, given any function $g(z), z \in C$, which is analytic outside the spectrum of L, the matrix $g(L)$ can be defined by the Cauchy formula

$$g(L) = \frac{1}{2\pi i} \int_\Gamma \frac{g(z)}{z - L} \, dz, \tag{8.2.2}$$

where, of course, $1/(z - L)$ is understood as inverse to the operator $z - L$ and the integration is performed in the counterclockwise direction along any smooth curve surrounding the spectrum. The curve Γ need not be closed provided that by Jordan's lemma the integral can be transformed into an integral along a closed curve. This gives us a lead to defining e^{tL} in the case of elliptic operators, especially since from Theorem 4.4.2 we know about the existence of the inverse operator for $z - L$ for a major part of the complex plane.

Thus, for $\eta \in (\pi/2, \pi)$ as in Sec. 4.4 define

$$E = E_\eta = \{z = a + ib \in C \setminus \{0\} : |\arg z| \leq \eta\}.$$

Also take $z_0 = z_0(\eta)$ from Theorem 4.4.2 and the operators $\mathcal{R}_z$ from Remark 4.4.4 defined for $z \in \cup_{\pi/2<\eta<\pi}(E_\eta \setminus B_{z_0(\eta)})$.

THEOREM 8.2.1. *For* $\eta \in (\pi/2, \pi)$ *and*

$$f \in C^\delta(\mathbb{R}^d), \quad t \in \Theta_\eta := \{t = a + ib \in C \setminus \{0\} : |\arg t| < \eta - \pi/2\}, \quad x \in \mathbb{R}^d$$

define

$$u(t,x) = T_t f(x) = \frac{1}{2\pi i} \int_{\partial(E_\eta \setminus B_{z_0})} e^{tz} \mathcal{R}_z f(x)\, dz, \tag{8.2.3}$$

where the integral is taken in the counterclockwise direction. Then

(i) for any $x \in \mathbb{R}^d$ *the function* $u(t,x)$ *as a function of* t *is analytic in the sector* Θ_η*; in particular, it is infinitely differentiable in* Θ_η*;*

(ii) for any $t \in \Theta_\eta$ *we have* $u(t,\cdot) \in C^{2+\delta}(\mathbb{R}^d)$ *and*

$$|u(t,\cdot)|_{2+\delta} \leq N|f|_\delta(|t|^{-1-\delta/2} + e^{|t|z_0}), \tag{8.2.4}$$

where $N = N(\kappa, K, \delta, \eta, d)$*;*

(iii) we have

$$\frac{\partial u}{\partial t} = Lu \quad in \quad \Theta_\eta \times \mathbb{R}^d;$$

(iv) for t *real we have*

$$|T_t f|_0 \leq |f|_0, \quad \lim_{t>0, t\to 0} |u(t,\cdot) - f|_0 = 0. \tag{8.2.5}$$

REMARK 8.2.2. It follows from the proof below that $u(t,x)$ is independent of η provided that $t \in \Theta_\eta$. Therefore, the arbitrariness of η shows that the function $u(t,x)$ defined by (8.2.3) is an analytic function of t for all $\operatorname{Re} t > 0$.

EXERCISE 8.2.3. Let A be a $d \times d$ matrix. Assume that all eigenvalues of A belong to B_{z_0} and take $(z - A)^{-1}$ instead of $\mathcal{R}_z$ in (8.2.3). Show that for $t > 0$ and any d-vector f in place of $f(x)$ formula (8.2.3) gives $[\exp(tA)]f$.

EXERCISE 8.2.4. In the situation of Exercise 8.2.3 we have

$$e^{tA} = \frac{1}{2\pi i} \int_{\partial B_{z_0}} e^{tz} (z - A)^{-1}\, dz.$$

By differentiating this formula with respect to t prove (8.2.2) for polynomials g. By plugging $g(z) = \det(z - A)$ and remembering that $g(z)(z - A)^{-1}$ is a polynomial prove the Hamilton-Kelly theorem: $g(A) = 0$.

EXERCISE 8.2.5. Given a $d \times d$ matrix A, prove that for positive z large enough

$$(z - A)^{-1} = \int_0^\infty e^{tA} e^{-zt}\, dt.$$

EXERCISE 8.2.6. Given a $d \times d$ matrix A and sufficiently large $z > 0$, *define*

$$(z - A)^{1/2} = c \int_0^\infty \frac{1}{t^{3/2}} \left\{ e^{At} e^{-zt} - 1 \right\} dt,$$

where c is a constant. Prove that $(z-A)^{1/2}(z-A)^{1/2} = z - A$ for an appropriate c.

8.3. Proof of Theorem 8.2.1

By Remark 4.4.4 we have $|\mathcal{R}_z f(x)| \leq N|z|^{-1+\delta/2}$ in $E_\eta \setminus B_{z_0}$ (f and x are fixed). Further, as is easy to see for any $\varepsilon \in (0, \eta - \pi/2)$ and $t \in \Theta_{\eta-\varepsilon}$, $z \in \partial E_\eta$, we have

$$|e^{tz}| = e^{\mathrm{Re}\,(tz)} \leq e^{-|t|\cdot|z| \sin \varepsilon}$$

and $|e^{tz}| \leq e^{|t|z_0}$ on ∂B_{z_0}. Also the function $\mathcal{R}_z f(x)$ is continuous (analytic by Exercise 4.4.5) in $E_\eta \setminus B_{z_0}$. It follows that the integral in (8.2.3) exists, and one can differentiate it with respect to $t \in \Theta_{\eta-\varepsilon}$. Therefore, it is analytic in $\Theta_{\eta-\varepsilon}$ and hence in Θ_η. We have proved (i). Also if $t \in \Theta_\eta$, we get

$$\frac{\partial u}{\partial t}(t,x) = \frac{1}{2\pi i} \int_{\partial(E_\eta \setminus B_{z_0})} e^{tz} z \mathcal{R}_z f(x)\, dz. \tag{8.3.1}$$

Next, we need some more information from Exercise 4.4.5. At first, notice that the theory of the Riemann integral of Banach space-valued continuous functions is quite similar to the theory of the Riemann integral for real–valued functions. By Exercise 4.4.5 and Remark 4.4.4 the $C^{2+\delta}(\mathbb{R}^d)$–valued function $\mathcal{R}_z f$ is continuous and bounded in $E_\eta \setminus B_{z_0}$. Therefore, for any $R \in (0, \infty)$ the integral

$$I_R := \frac{1}{2\pi i} \int_{\partial(E_\eta \setminus B_{z_0}), |z| \leq R} e^{tz} \mathcal{R}_z f\, dz$$

is defined as the $C^{2+\delta}(\mathbb{R}^d)$–limit of integral sums, and

$$\frac{1}{2\pi i} \int_{\partial(E_\eta \setminus B_{z_0})} e^{tz} \mathcal{R}_z f\, dz \tag{8.3.2}$$

is defined as the $C^{2+\delta}(\mathbb{R}^d)$–limit of I_R. Convergence in $C^{2+\delta}(\mathbb{R}^d)$, in particular, implies pointwise (actually uniform) convergence of functions along with their derivatives of order up to and including 2. This means that $u(t,x)$ is the value at x of the $C^{2+\delta}(\mathbb{R}^d)$ function (8.3.2), so that we have the first assertion in (ii).

Further, for $t \in \Theta_{\eta-\varepsilon}$ by Remark 4.4.4

$$|u(t,\cdot)|_{2+\delta} \leq \left| \frac{1}{2\pi i} \int_{\partial(E_\eta \setminus B_{z_0})} e^{tz} \mathcal{R}_z f\, dz \right|_{2+\delta} \leq N|f|_\delta \int_{\partial(E_\eta \setminus B_{z_0})} |e^{tz}| \cdot |z|^{\delta/2}\, dl$$

$$\leq N|f|_\delta \Big(\int_{\partial E_\eta} e^{-|t|\cdot|z| \sin \varepsilon} |z|^{\delta/2}\, dl + \int_{\partial B_{z_0}} e^{|t|\cdot|z|} |z|^{\delta/2}\, dl \Big)$$

$$= N|f|_\delta (N_1(|t| \sin \varepsilon)^{-1-\delta/2} + N_2 e^{|t|z_0}).$$

We thus get (8.2.4), but only for $t \in \Theta_{\eta-\varepsilon}$. However, owing to the analyticity of $\mathcal{R}_z$ we could replace η in (8.2.3) with any bigger number from $(\pi/2, \pi)$ without affecting the result of integration.

Turning our attention to (iii), notice that in (8.3.1) by definition $z\mathcal{R}_z f = L\mathcal{R}_z f(x) + f$ and by Jordan's lemma

$$f(x)\int_{\partial(E_\eta\setminus B_{z_0})} e^{tz}\,dz = f(x)\int_{\partial B_{z_0}} e^{tz}\,dz = 0,$$

so that

$$\frac{\partial u}{\partial t}(t,x) = \frac{1}{2\pi i}\int_{\partial(E_\eta\setminus B_{z_0})} e^{tz} L\mathcal{R}_z f(x)\,dz.$$

As above, this time considering the integral of $C^\delta(\mathbb{R}^d)$-valued function $L\mathcal{R}_z f$ and using that for any finite integral sum the operator L can be pulled out, one shows that one can pull the operator L out in the last formula, and in this way one obtains assertion (iii).

Next, we prove the second relation in (8.2.5) under the additional assumption that $f \in C^{2+\delta}(\mathbb{R}^d)$. Use again Jordan's lemma to prove that

$$\frac{1}{2\pi i}\int_{\partial(E_\eta\setminus B_{z_0})} e^{tz} z^{-1}\,dz = \frac{1}{2\pi i}\int_{\partial B_{z_0}} e^{tz} z^{-1}\,dz = 1,$$

and for $n > 0$

$$\int_{\partial(E_\eta\setminus B_{z_0})} z^{-1}\mathcal{R}_z Lf(x)\,dz = \int_{\operatorname{Re} z = z_0+n} z^{-1}\mathcal{R}_z Lf(x)\,dz$$

$$= \lim_{n\to\infty}\int_{\operatorname{Re} z = z_0+n} z^{-1}\mathcal{R}_z Lf(x)\,dz = 0.$$

Therefore,

$$u(t,x) - f(x) = \frac{1}{2\pi i}\int_{\partial(E_\eta\setminus B_{z_0})} e^{tz}[\mathcal{R}_z f(x) - z^{-1} f(x)]\,dz$$

$$= \frac{1}{2\pi i}\int_{\partial(E_\eta\setminus B_{z_0})} e^{tz} z^{-1}\mathcal{R}_z Lf(x)\,dz = \frac{1}{2\pi i}\int_{\partial(E_\eta\setminus B_{z_0})} (e^{tz}-1) z^{-1}\mathcal{R}_z Lf(x)\,dz.$$

Hence owing to the estimate $|\mathcal{R}_z f|_\delta \le N|z|^{-1+\delta/2}|f|_\delta$ and the dominated convergence theorem, we have

$$|u(t,\cdot) - f|_\delta \le N|Lf|_\delta \int_{\partial(E_\eta\setminus B_{z_0})} |e^{tz}-1|\cdot|z|^{-2+\delta/2}\,dl \to 0$$

as $\Theta_\eta \ni t \to 0$, in particular, as $t \downarrow 0$. We have proved the second assertion in (8.2.5) even for a higher norm if $f \in C^{2+\delta}(\mathbb{R}^d)$.

This and assertions (i) through (iii) show that u is a bounded solution to the Cauchy problem (8.2.1). By the maximum principle applied to $\operatorname{Re}(\xi u)$, where ξ is any fixed unitary complex number, we get $|u(t,\cdot)|_0 \le |f|_0$ or $|T_t f|_0 \le |f|_0$ if $f \in C^{2+\delta}(\mathbb{R}^d)$. Thus we have proved assertion (iv) for $f \in C^{2+\delta}(\mathbb{R}^d)$.

To obtain it in full generality, take arbitrary $f \in C^\delta(\mathbb{R}^d)$ and find functions $f_n \in C^{2+\delta}(\mathbb{R}^d)$ such that $|f - f_n|_\sigma \to 0$ as $n \to \infty$, where $\sigma \in (0, \delta)$. Then for any $t > 0$ we have $|T_t f - T_t f_n|_{2+\sigma} \le N|f - f_n|_\sigma$, $|T_t f - T_t f_n|_0 \to 0$,

$$|T_t f|_0 \le \limsup_{n\to\infty} |T_t f_n|_0 \le \limsup_{n\to\infty} |f_n|_0 = |f|_0,$$

and the first assertion in (8.2.5) is proved. Also

$$\limsup_{t\downarrow 0} |T_t f - f|_0 \le \limsup_{t\downarrow 0} |T_t f_n - f_n|_0 + 2|f - f_n|_0 = 2|f - f_n|_0$$

for any n, and by letting $n \to \infty$ we get the second assertion in (8.2.5) in its full generality as well. The theorem is proved.

COROLLARY 8.3.1. *Given $f \in C^\delta(\mathbb{R}^d)$ there exists a unique bounded solution of the Cauchy problem (8.2.1). This solution is given by $T_t f(x)$.*

REMARK 8.3.2. We were talking about the impossibility of defining $L^n f$. Interestingly enough $u(t, x)$ approximates $f(x)$, and for any $f \in C^\delta(\mathbb{R}^d)$ and any $t > 0$ we have $Lu(t, \cdot) \in C^{2+\delta}(\mathbb{R}^d)$, $L^2 u = L(Lu)(t, \cdot) \in C^{2+\delta}(\mathbb{R}^d)$ and so on. Indeed,

$$Lu(t, x) = \frac{1}{2\pi i} \int_{\partial(E_\eta \setminus B_{z_0})} e^{tz} L\mathcal{R}_z f\, dz = \frac{1}{2\pi i} \int_{\partial(E_\eta \setminus B_{z_0})} e^{tz} z\mathcal{R}_z f\, dz,$$

$$L^2 u(t, x) = \frac{1}{2\pi i} \int_{\partial(E_\eta \setminus B_{z_0})} e^{tz} zL\mathcal{R}_z f\, dz = \frac{1}{2\pi i} \int_{\partial(E_\eta \setminus B_{z_0})} e^{tz} z^2\mathcal{R}_z f\, dz,$$

and so on.

EXERCISE 8.3.3. For $t \in \Theta_\eta$ prove that $|T_t f|_\delta \le N|f|_\delta(|t|^{-\delta/2} + \exp(|t|z_0)$.

It is worth mentioning that $|t|^{-\delta/2}$ in the last estimate can be replaced by 1 at least if t is real.

EXERCISE 8.3.4. By using the definition of $\mathcal{R}_z$ and the properties of T_t prove that for $z > 0$, $f \in C^\delta(\mathbb{R}^d)$, $x \in \mathbb{R}^d$ we have

$$\mathcal{R}_z f(x) = \int_0^\infty e^{-tz} T_t f(x)\, dt, \tag{8.3.3}$$

which means that the result of Exercise 8.2.5 holds true for $A = \Delta$. By using the unique continuation property and/or Exercise 4.4.6 prove that (8.3.3) also holds for $\operatorname{Re} z > 0$.

EXERCISE 8.3.5. By using the maximum principle show that if $f \in C^\delta(\mathbb{R}^d)$ and $0 \le f \le 1$, then $0 \le T_t f \le 1$ for $t > 0$.

EXERCISE 8.3.6 (semigroup property). Prove that for $t, s > 0$ and $f \in C^\delta(\mathbb{R}^d)$ we have $T_t T_s f = T_{t+s} f$.

8.4. The heat equation

For $z = (t, x) \in \mathbb{R}^{d+1} = \{z = (t, x) : t \in (-\infty, \infty), x \in \mathbb{R}^d\}$ and constant $\lambda \geq 0$ define

$$G_\lambda(z) = G_\lambda(t, x) = \begin{cases} \frac{1}{(4\pi t)^{d/2}} \exp(-\frac{1}{4t}|x|^2 - \lambda t) & \text{if} \quad t > 0, \\ 0 & \text{if} \quad t \leq 0. \end{cases} \tag{8.4.1}$$

We will use the fact that the function $G_\lambda(z)$ is infinitely differentiable with respect to z for $z \neq 0$ and each of its derivatives, of any order, is bounded outside any neighborhood of the origin. Also we will use the fact that for $t > 0$ and $x \in \mathbb{R}^d$

$$\Delta G_\lambda(t, x) - \frac{\partial}{\partial t} G_\lambda(t, x) - \lambda G_\lambda(t, x) = 0, \quad \int_{\mathbb{R}^d} G_\lambda(t, x)\, dx = e^{-\lambda t}. \tag{8.4.2}$$

The last equality shows that for $\lambda > 0$ convolutions with respect to (t, x) of G_λ with bounded functions are bounded. This is also true for $\lambda = 0$ if the functions have compact support with respect to t. The first equation in (8.4.2) and the rules of differentiation of integrals imply the following result.

LEMMA 8.4.1. *Let $\lambda > 0$ and Q be a domain in $\mathbb{R}^{d+1}$ and f be a bounded function vanishing in Q. Then the function*

$$u(t, x) := G_\lambda * f(t, x) = \int_{-\infty}^{\infty} \int_{\mathbb{R}^d} G_\lambda(t - s, x - y) f(s, y)\, dyds$$

is infinitely differentiable in Q and satisfies $\Delta u - u_t - \lambda u = 0$ in Q. These assertions hold true for $\lambda = 0$ as well if in addition there is a constant $T < \infty$ such that $f(t, x) = 0$ for any $|t| \geq T$ and x.

The function G_λ is called the fundamental solution of the heat equation

$$\Delta u - u_t - \lambda u + f = 0 \quad \text{in} \quad \mathbb{R}^{d+1}. \tag{8.4.3}$$

Actually, usually one calls the heat equation equation (8.4.3) with $\lambda = 0$. It is worth noticing that equation (8.4.3) is easier to investigate, and if u is its solution, then $ue^{\lambda t}$ is a solution of (8.4.3) with $\lambda = 0$ and with $fe^{\lambda t}$ instead of f.

Upon observing that $G_\lambda * f(s, x)$ coincides with $\mathcal{R}f(s, x)$ introduced in (2.8.5) when a is the $d \times d$ identity matrix and $b = 0$ and $c = -\lambda$, from Theorem 2.8.3 we obtain the following result for $\lambda > 0$.

THEOREM 8.4.2. *Let $-\infty < S < T < \infty$.*

(i) If $u \in C^2(\mathbb{R}^{d+1})$ and $u(t, x) = 0$ for $t \notin [S, T]$, then in $\mathbb{R}^{d+1}$ we have $u(z) = -G_\lambda(z) * (\Delta u - u_t - \lambda u)(z)$.

(ii) If $f \in C^2(\mathbb{R}^{d+1})$ and $f(t, x) = 0$ for $t \notin [S, T]$ and $u := -G_\lambda * f$, then $u \in C^2(\mathbb{R}^{d+1})$ and u satisfies (8.4.3).

To get these assertions for $\lambda = 0$, it suffices to make the passage to the limit as $\lambda \downarrow 0$ in the equations

$$u(z) = -G_\lambda(z) * (\Delta u - u_t - \lambda u)(z), \quad f(z) = -G_\lambda(z) * (\Delta f - f_t - \lambda f)(z),$$

observing that the second equality in (8.4.2) and boundedness with respect to t of the supports of u, f allow us to make this passage to the limit.

REMARK 8.4.3. Assertion (i) of Theorem 8.4.2 means that in the sense of distributions $\Delta G_\lambda - G_{\lambda t} - \lambda G_\lambda = -\delta_0$ in $\mathbb{R}^{d+1}$, which explains why G is called the fundamental solution of (8.4.3). The method of finding G_λ is to make the Fourier transform with respect to x in the first equation in (8.4.2). Then we see that $\tilde{G}_\lambda(t,\xi)$ should satisfy the equation $\tilde{G}_{\lambda t} = -(|\xi|^2+\lambda)\tilde{G}_\lambda$. One of its solutions is $\exp(-(|\xi|^2+\lambda)t)$, and then one recovers G_λ by applying the inverse Fourier transform.

Next we generalize Theorem 2.5.2 for functions depending on t. Denote $D_t = \partial/\partial t$ and $D_x^\alpha = D^\alpha$.

THEOREM 8.4.4. *Let $R>0$, $Q_R := (-R^2,0)\times B_R$, $u\in C(\bar{Q}_R)$, u be infinitely differentiable in Q_R and $\Delta u - u_t = 0$ in Q_R. Then for any multi-index $\alpha\in\mathbb{R}^d$ and integer $n\geq 0$ we have*

$$|D_t^n D_x^\alpha u(0)| \leq \frac{N^{|\alpha|+2n}(d)(|\alpha|+2n)^{|\alpha|+2n}}{R^{|\alpha|+2n}}|u|_{0;Q_R}. \tag{8.4.4}$$

Proof. First notice that the equation $\Delta u - u_t = 0$ remains invariant under the change of $u(t,x)$ by $u(R^2t, Rx)$. It follows that we need to prove (8.4.4) only for $R=1$.

We will apply Bernstein's method. Take any $\zeta\in C_0^\infty(\mathbb{R}^{d+1})$ with support in $(-R^2,R^2)\times B_R$, assume that $\zeta(0)=1$ and consider the function

$$w := \zeta^2|\mathrm{grad}_x\, u|^2 + \mu|u|^2,$$

where $\mu>0$ is a constant. We have $\Delta u - u_t = 0$, $\Delta u_{x^i} - u_{x^it} = 0$, and

$$\Delta w - w_t = |\mathrm{grad}_x\, u|^2\Delta(\zeta^2) + \zeta^2[2u_{x^i}\Delta u_{x^i} + 2\sum_{ij} u^2_{x^ix^j}] + 8\zeta\zeta_{x^i}u_{x^j}u_{x^ix^j}$$

$$+2\mu|\mathrm{grad}_x\, u|^2 + 2\mu u\Delta u - 2\zeta\zeta_t|\mathrm{grad}_x\, u|^2 - 2\zeta^2u_{x^i}u_{x^it} - 2\mu uu_t$$

$$= |\mathrm{grad}_x\, u|^2[2\mu + \Delta(\zeta^2) - 2\zeta\zeta_t] + 2\zeta^2\sum_{ij}u^2_{x^ix^j} + 8[\zeta_{x^i}u_{x^j}]\cdot[\zeta u_{x^ix^j}]$$

$$\geq |\mathrm{grad}_x\, u|^2[2\mu + \Delta(\zeta^2) - 8|\mathrm{grad}_x\zeta|^2 - 2\zeta\zeta_t].$$

We see how to take μ so that $\Delta w - w_t \geq 0$. Fix such a μ. Then by the maximum principle

$$|\mathrm{grad}_x\, u|^2(0) \leq \sup_{Q_1} w \leq \sup_{\partial' Q_1} w = \mu \sup_{\partial' Q_1}|u|^2.$$

This yields (8.4.4) for $|\alpha|=1$, $n=0$ and $R=1$, and hence for all $R>0$. To prove it for $|\alpha|=2$ and still $n=0$, observe that derivatives of u also satisfy the homogeneous heat equation. Therefore,

$$|D_iD_ju(0)| \leq \frac{N(d)}{R/2}|D_iu|_{0;Q_{R/2}} \leq \frac{N(d)}{R/2}\frac{N(d)}{R/2}|u|_{0;Q_R}.$$

We can proceed like this for any $|\alpha|$ so that (8.4.4) is proved for $n=0$. If $n\geq 1$, it suffices to notice that $D_tD^\alpha u=\Delta D^\alpha u$,..., $D_t^nD^\alpha u=\Delta^nD^\alpha u$, so that one gets (8.4.4) for $n\geq 1$ from the previous case. The theorem is proved.

EXERCISE 8.4.5 (cf. Corollary 2.5.3). Let Q be a domain in $\mathbb{R}^{d+1}$ and $\{u_n\}$ be a sequence of functions infinitely differentiable in Q. Assume that u_n are uniformly bounded and $\Delta u_n-u_{nt}=0$ in Q. Prove that for any multi–index α and integer k the family $D_t^kD_x^\alpha u_n$ is uniformly bounded and equicontinuous on any compact set $\Gamma\subset Q$ and if $u_n(z)\to u(z)$ at any point $z\in Q$, then u is infinitely differentiable in Q and $\Delta u-u_t=0$ in Q.

EXERCISE 8.4.6 (cf. Corollary 2.5.4). Let u be infinitely differentiable in $\mathbb{R}^{d+1}$ and $\Delta u-u_t=0$ in $\mathbb{R}^{d+1}$. Assume that $|u(z)|\leq N(1+|z|^n)$ for some constants N,n and all z. Prove that u is a polynomial in z. Also prove (Liouville's theorem) that if $n=0$ so that u is bounded, then u is constant.

Observe that in the parabolic case Liouville's theorem is no longer true for nonnegative (unbounded) functions. The counterexample is given by the function $\exp(x^i+t)$.

EXERCISE 8.4.7. Prove that if u is infinitely differentiable in Q for a domain $Q\subset\mathbb{R}^{d+1}$ and $\Delta u-u_t=0$ in Q, then for any t the function $u(t,x)$ is real–analytic in $Q(t)=\{x:(t,x)\in Q\}$. Conclude that if $u(t,x)=0$ in a subdomain Ω of $Q(t)$, then $u(t,\cdot)\equiv 0$ in any connected subdomain of $Q(t)$ containing Ω.

Interestingly enough, in the situation of this exercise u need not be real–analytic in t; in particular, it might happen that $u(t,x)=0$ for $(t,x)\in\{(s,y)\in Q:s\leq 1\}$ but $u\not\equiv 0$ (cf. Remark 10.3.4).

EXERCISE* 8.4.8. Prove that if $\lambda\geq 0$ and $R>0$, then $|G_\lambda * I_{Q_R}|\leq NR^2$, where $N=N(d)$.

8.5. Parabolic Hölder spaces

In $\mathbb{R}^{d+1}$ define the parabolic distance between the points $z_1=(t_1,x_1)$, $z_2=(t_2,x_2)$ as

$$\rho(z_1,z_2)=|x_1-x_2|+|t_1-t_2|^{1/2}.$$

As always we fix a constant $\delta\in(0,1)$, although the construction of this section is valid for $\delta=1$ as well.

EXERCISE* 8.5.1. Prove the triangle inequality for $\rho(z_1,z_2)$.

If u is a function in a domain $Q\subset\mathbb{R}^{d+1}$, we denote

$$[u]_{\delta/2,\delta;Q}=\sup_{\substack{z_1\neq z_2\\ z_i\in Q}}\frac{|u(z_1)-u(z_2)|}{\rho^\delta(z_1,z_2)},\qquad |u|_{\delta/2,\delta;Q}=|u|_{0;Q}+[u]_{\delta/2,\delta;Q}.$$

By $C^{\delta/2,\delta}(Q)$ we denote the space of all functions u for which $|u|_{\delta/2,\delta;Q}<\infty$. We also introduce the parabolic Hölder space $C^{1+\delta/2,2+\delta}(Q)$ as the set of all real–valued functions $u(z)$ defined in Q for which both

$$[u]_{1+\delta/2,2+\delta;Q}:=[u_t]_{\delta/2,\delta;Q}+\sum_{i,j=1}^d[u_{x^ix^j}]_{\delta/2,\delta;Q}<\infty,$$

$$|u|_{1+\delta/2,2+\delta;Q} := |u|_{0;Q} + |u_x|_{0;Q} + |u_t|_{0;Q} + \sum_{i,j=1}^{d} |u_{x^ix^j}|_{0;Q} + [u]_{1+\delta/2,2+\delta;Q} < \infty.$$

As in Remark 3.1.3 one can show that $C^{\delta/2,\delta}(Q)$ and $C^{1+\delta/2,2+\delta}(Q)$ are Banach spaces.

Notice that if the domain Q is not too bad (for example, convex), then any function $C^{1+\delta/2,2+\delta}(Q)$-function is uniformly continuous in Q and admits a unique continuous extension on $\bar{Q}$. Therefore, without ambiguity we can speak about values of a function $u \in C^{1+\delta/2,2+\delta}(Q)$ on ∂Q. As in the case of elliptic equations we drop the subscript Q if $Q = \mathbb{R}^{d+1}$.

Also notice that from the elementary inequality

$$|u(z_1)v(z_1) - u(z_2)v(z_2)| \leq |u(z_1)| \cdot |v(z_1) - v(z_2)| + |v(z_2)| \cdot |u(z_1) - u(z_2)|$$

we obtain the following inequality, which will be constantly in use:

$$[uv]_{\delta/2,\delta;Q} \leq |u|_{0;Q} \cdot [v]_{\delta/2,\delta;Q} + |v|_{0;Q} \cdot [u]_{\delta/2,\delta;Q} \quad \forall u, v \in C^{\delta/2,\delta}(Q).$$

We also define other seminorms based on approximations. Let $\mathcal{P}_2$ be the set of all polynomials of the variables $t, x^1, ..., x^d$ of the type

$$\alpha t + \alpha^i x^i + \alpha^{ij} x^i x^j + \beta.$$

Recall that $B_\rho(x) = \{y \in \mathbb{R}^d : |x - y| < \rho\}$, and for $z = (t, x) \in \mathbb{R}^{d+1}$ let $Q_\rho(z) = (t - \rho^2, t) \times B_\rho(x) = z + Q_\rho$ and

$$[u]'_{1+\delta/2,2+\delta} = \sup_{z\in\mathbb{R}^{d+1}} \sup_{\rho>0} \frac{1}{\rho^{2+\delta}} \inf_{p\in\mathcal{P}_2} |u - p|_{0;Q_\rho(z)}.$$

THEOREM 8.5.2. *There exists a constant $N = N(d)$ such that for any $u \in C^{1+\delta/2,2+\delta}(\mathbb{R}^{d+1})$ we have*

$$[u]'_{1+\delta/2,2+\delta} \leq N[u]_{1+\delta/2,2+\delta}, \quad [u]_{1+\delta/2,2+\delta} \leq N[u]'_{1+\delta/2,2+\delta}. \tag{8.5.1}$$

Proof. To prove the first inequality we observe that by Taylor's formula for any fixed $z_0 = (t_0, x_0) \in \mathbb{R}^{d+1}$ and $z = (t, x)$

$$u(z) = u(t_0, x) + (t - t_0)u_t(\theta, x) = u(z_0) + (t - t_0)u_t(\theta, x)$$

$$+u_{x^i}(z_0)(x^i - x_0^i) + \frac{1}{2} u_{x^ix^j}(t_0, \xi)(x^i - x_0^i)(x^j - x_0^j),$$

where θ lies in the interval with the ends t, t_0 and $\xi \in [x, x_0]$. Hence for Taylor's polynomial

$$T_{z_0}u(z) := u(z_0) + (t - t_0)u_t(z_0) + u_{x^i}(z_0)(x^i - x_0^i) + \frac{1}{2} u_{x^ix^j}(z_0)(x^i - x_0^i)(x^j - x_0^j) \tag{8.5.2}$$

and $\rho(z, z_0) \leq \rho$, we have ($|t - t_0| \leq \sqrt{\rho}, |x - x_0| \leq \rho$, and)

$$|u(z) - T_{z_0}u(z)| \leq \rho^2|u_t(\theta, x) - u_t(z_0)| + \rho^2 \sum_{i,j=1}^{d} |u_{x^ix^j}(t_0, \xi) - u_{x^ix^j}(z_0)|$$

$$\leq N\rho^2[u]_{1+\delta/2,2+\delta}(\rho^\delta((\theta, x), z_0) + \rho^\delta((t_0, \xi), z_0)) \leq N\rho^{2+\delta}[u]_{1+\delta/2,2+\delta}. \qquad (8.5.3)$$

This proves the first inequality in (8.5.1).

To prove the second one, we denote by D one of the operators

$$\frac{\partial}{\partial t}, \quad \frac{\partial^2}{\partial x^i \partial x^j}$$

and associate with them, respectively, the finite–difference operators σ_h, $h > 0$,

$$u \longrightarrow \frac{1}{h^2}[u(t, x) - u(t - h^2, x)],$$

$$u \longrightarrow \frac{1}{h^2}[u(t, x + he_i + he_j) - u(t, x + he_i) - u(t, x + he_j) + u(t, x)]. \qquad (8.5.4)$$

Also, let σ'_h be the operators defined by, respectively,

$$u \to u_t(t - h^2, x), \quad u \to \frac{1}{2}[u_{x^ix^i}(t, x + he_i + he_j) - u_{x^ix^i}(t, x + he_i)$$

$$+ u_{x^jx^j}(t, x + he_i + he_j) - u_{x^jx^j}(t, x + he_j) + 2u_{x^ix^j}(t, x + he_i + he_j)].$$

By considering the expressions in the brackets in the right–hand sides in (8.5.4) as functions of h and using Taylor's formula, we get

$$\sigma_h u(z) = \sigma'_{h'} u(z),$$

where $h' \leq h$ and h' depends on z. It follows, in particular, that for any $p \in \mathcal{P}_2$ the expression $\sigma_h p$ is a constant independent of h, z. Also, if σ_h corresponds to D and $u \in C^{1+\delta/2,2+\delta}(\mathbb{R}^{d+1})$, then

$$|\sigma_h u(z) - Du(z)| = |\sigma'_{h'} u(z) - Du(z)| \leq Nh^\delta[u]_{1+\delta/2,2+\delta},$$

where N is an absolute constant. Now take any z_1, z_2, denote $\rho = \rho(z_1, z_2)$ and take $h = \varepsilon\rho$, where the constant $\varepsilon \in (0, 1)$ will be specified later. Without loss of generality we assume that $t_1 \leq t_2$. Then all points $(t_n - h^2, x_n)$, $(t_n, x_n + he_i + he_j)$, $n = 1, 2$, belong to $Q_{3\rho}(z_2)$. Hence for any $p \in \mathcal{P}_2$ we have

$$|Du(z_1) - Du(z_2)| \leq |Du(z_1) - \sigma_h u(z_1)| + |Du(z_2) - \sigma_h u(z_2)|$$

$$+ |\sigma_h(u - p)(z_1) - \sigma_h(u - p)(z_2)| \leq Nh^\delta[u]_{1+\delta/2,2+\delta} + |\sigma_h(u - p)(z_1)|$$

$$+ |\sigma_h(u - p)(z_2)|, \quad |\sigma_h(u - p)(z_i)| \leq \frac{4}{h^2}|u - p|_{0;Q_{3\rho}(z_2)}.$$

Since this is true for any $p \in \mathcal{P}_2$, we have

$$|Du(z_1) - Du(z_2)| \leq N\varepsilon^\delta\rho^\delta[u]_{1+\delta/2,2+\delta} + N\varepsilon^{-2}\rho^\delta[u]'_{1+\delta/2,2+\delta},$$

$$[u]_{1+\delta/2,2+\delta} \le N_1 \varepsilon^\delta [u]_{1+\delta/2,2+\delta} + N\varepsilon^{-2}[u]'_{1+\delta/2,2+\delta},$$

where the last inequality follows from the previous one after dividing through by ρ^δ, taking the least upper bound of the left–hand side with respect to z_i and summing up with respect to different operators D. It remains only to choose ε so that $N_1\varepsilon^\delta \le 1/2$. The theorem is proved.

REMARK 8.5.3. If u is independent of t, the theorem gives Theorem 3.3.1 with $k=2$.

EXERCISE 8.5.4. Take Taylor's polynomials $T_{z_0}u(z)$ from (8.5.2) and define

$$[u]''_{1+\delta/2,2+\delta} = \sup_{z_0\in\mathbb{R}^{d+1}} \sup_{\rho>0} \rho^{-2-\delta} \sup_{z\in Q_\rho(z_0)} |u(z) - T_{z_0}u(z)|.$$

Prove that all three seminorms $[\cdot]_{1+\delta/2,2+\delta}$, $[\cdot]'_{1+\delta/2,2+\delta}$, $[\cdot]''_{1+\delta/2,2+\delta}$ are equivalent on $C^{1+\delta/2,2+\delta}(\mathbb{R}^{d+1})$.

EXERCISE* 8.5.5. Define

$$[u]'_{\delta/2,\delta;Q} = \sup_{(t,x),(s,x)\in Q} \frac{|u(t,x)-u(s,x)|}{|t-s|^{\delta/2}} + \sup_{(t,x),(t,y)\in Q} \frac{|u(t,x)-u(t,y)|}{|x-y|^\delta},$$

and prove that for convex domains Q or for cylindrical Q the seminorms $[\cdot]'_{\delta/2,\delta;Q}$ and $[\cdot]_{\delta/2,\delta;Q}$ are equivalent.

EXERCISE* 8.5.6. Prove that if we are given a sequence of $u_n \in C^{1+\delta/2,2+\delta}(Q)$ such that $|u_n|_{1+\delta/2,2+\delta;Q}$ is bounded, then there is a subsequence $\{v_n\}$ of $\{u_n\}$ and a function $u \in C^{1+\delta/2,2+\delta}(Q)$ such that $v_n, v_{nx}, v_{nt}, v_{nxx}$ converge to u, u_x, u_t, u_{xx} in the sense of $C(\Gamma)$ whenever compact $\Gamma \subset Q$ and also

$$|u|_{1+\delta/2,2+\delta;Q} \le \liminf |v_n|_{1+\delta/2,2+\delta;Q}.$$

EXERCISE* 8.5.7. Let $\zeta \in C_0^\infty(\mathbb{R}^{d+1}), \zeta \ge 0$ and $\int \zeta\, dz = 1$. For $\varepsilon > 0$ define $\zeta_\varepsilon(z) = \varepsilon^{-d-1}\zeta(z\varepsilon^{-1})$ and let $u^\varepsilon := \zeta_\varepsilon * u$. Prove that

(i) for any bounded measurable u we have $u^\varepsilon \in C_b^\infty(\mathbb{R}^{d+1})$;

(ii) if $u \in C^{1+\delta/2,2+\delta}(\mathbb{R}^{d+1})$, then

$$|u^\varepsilon|_{1+\delta/2,2+\delta} \le |u|_{1+\delta/2,2+\delta}, \quad |u|_{1+\delta/2,2+\delta} = \lim_{\varepsilon\downarrow 0} |u^\varepsilon|_{1+\delta/2,2+\delta}.$$

EXERCISE 8.5.8. Prove that the space $C^{\delta/2,\delta}([0,1]^2)$ is *not* separable.

EXERCISE* 8.5.9. Prove that if Q is convex, a subdomain $Q' \subset Q$, $u \in C^{1+\delta/2,2+\delta}(Q)$ and $u=0$ outside Q', then $|u|_{1+\delta/2,2+\delta;Q} = |u|_{1+\delta/2,2+\delta;Q'}$.

EXERCISE 8.5.10. Let $\zeta \in C_0^\infty(\mathbb{R}^{d+1}), \zeta \ge 0$ and $\int \zeta\, dz = 1$. For $\varepsilon > 0$ define $\zeta_\varepsilon(t,x) = \varepsilon^{-d-2}\zeta(t\varepsilon^{-2}, x\varepsilon^{-1})$ and let $u^{(\varepsilon)} := \zeta_\varepsilon * u$. For $0<\delta\le\gamma\le 1$ define

$$[u]''_{\delta/2,\delta} = \sup_{\varepsilon>0} \varepsilon^{\gamma-\delta}[u^{(\varepsilon)}]_{\gamma/2,\gamma}.$$

Prove that if $u \in C^{1+\delta/2,2+\delta}(\mathbb{R}^{d+1})$, then $[u]_{\delta/2,\delta} \le N(d,\delta,\gamma)[u]''_{\delta/2,\delta}$.

8.6. The basic a priori estimate

A priori estimates in Hölder–space norms and results about solvability of parabolic equations in Hölder spaces appeared much later than the corresponding results for elliptic equations. We mean the results due to Barrar (1956) and Friedman (1958). Below we follow the method of proof invented by Safonov about 1984.

Take and fix a constant $\delta \in (0,1)$.

THEOREM 8.6.1. *Let $u \in C_0^\infty(\mathbb{R}^{d+1})$. Define $f = \Delta u - u_t$. Then there exists a constant $N = N(d,\delta)$ such that*

$$[u]_{1+\delta/2,2+\delta} \le N[f]_{\delta/2,\delta}.$$

Proof. Take $z_0 \in \mathbb{R}^{d+1}$, $\rho > 0$ and a constant $K \ge 1$ which will be specified later. Also take $\zeta \in C_0^\infty(\mathbb{R}^{d+1})$ such that $\zeta = 1$ in $Q_{(K+1)\rho}(z_0)$. Taking the definition of Taylor's polynomials from (8.5.2), define

$$g = \Delta(\zeta T_{z_0}u) - (\zeta T_{z_0}u)_t.$$

In $Q_{(K+1)\rho}(z_0)$ we obviously have

$$g = \Delta(T_{z_0}u) - (T_{z_0}u)_t = f(z_0).$$

Therefore by Theorem 8.4.2 in $Q_{(K+1)\rho}(z_0)$ it holds that

$$u - T_{z_0}u = u - \zeta T_{z_0}u = -G_0 * (f-g) = G_0 * [(f(z_0) - f)I_{Q_{(K+1)\rho}(z_0)}]$$

$$+ G_0 * [(g-f)I_{Q^c_{(K+1)\rho}(z_0)}] =: r + h. \quad (8.6.1)$$

By Lemma 8.4.1 the function h is infinitely differentiable in $Q_{(K+1)\rho}(z_0)$ and $\Delta h - h_t = 0$ in $Q_{(K+1)\rho}(z_0)$. Also apply (8.5.3) to h instead of u and notice that for $z \in Q_\rho(z_0)$ we have

$$|h_t(\theta,x) - h_t(z_0)| \le |h_t(\theta,x) - h_t(t_0,x)| + |h_t(t_0,x) - h_t(t_0,x_0)|$$

$$\le \rho^2 |D_t^2 h|_{0;Q_\rho(z_0)} + \rho |\mathrm{grad}_x\, D_t h|_{0;Q_\rho(z_0)}.$$

Thus, from (8.5.3) and Theorem 8.4.4 we obtain

$$|h - T_{z_0}h|_{0;Q_\rho(z_0)} \le \rho^4 |D_t^2 h|_{0;Q_\rho(z_0)} + \rho^3 |\mathrm{grad}_x\, D_t h|_{0;Q_\rho(z_0)}$$

$$+\rho^3 \sum_{i,j,k=1}^d |D_iD_jD_k h|_{0;Q_\rho(z_0)} \le N(K^{-4} + K^{-3})|h|_{Q_{(K+1)\rho}(z_0)} \le NK^{-3}|h|_{Q_{(K+1)\rho}(z_0)}.$$

Furthermore, by Exercise 8.4.8 we have $|r| \le N(K+1)^{2+\delta}\rho^{2+\delta}[f]_{\delta/2,\delta}$, so that applying (8.5.3) again we find

$$|h|_{Q_{(K+1)\rho}(z_0)} = |u - T_{z_0}u - r|_{Q_{(K+1)\rho}(z_0)} \le |u - T_{z_0}u|_{Q_{(K+1)\rho}(z_0)}$$

$$+ N(K+1)^{2+\delta}\rho^{2+\delta}[f]_{\delta/2,\delta} \le N(K+1)^{2+\delta}\rho^{2+\delta}([u]_{1+\delta/2,2+\delta} + [f]_{\delta/2,\delta}).$$

This and (8.6.1) yield

$$\inf_{p\in\mathcal{P}_2} |u-p|_{0;Q_\rho(z_0)} \leq |u - T_{z_0}u - T_{z_0}h|_{0;Q_\rho(z_0)} \leq |h - T_{z_0}h|_{0;Q_\rho(z_0)} + |r|_{0;Q_\rho(z_0)}$$

$$\leq NK^{-3}(K+1)^{2+\delta}\rho^{2+\delta}[u]_{1+\delta/2,2+\delta} + N(K^{-3}+1)(K+1)^{2+\delta}\rho^{2+\delta}[f]_{\delta/2,\delta}.$$

Finally, we divide through by $\rho^{2+\delta}$ and take the least upper bound of the left–hand side of the last inequality with respect to all $z_0 \in \mathbb{R}^{d+1}$ and $\rho > 0$. Then owing to Theorem 8.5.2, we conclude

$$[u]_{1+\delta/2,2+\delta} \leq NK^{-3}(K+1)^{2+\delta}[u]_{1+\delta/2,2+\delta} + N(K+1)^{2+\delta}[f]_{\delta/2,\delta},$$

and it remains only to take K so large that the constant factor of $[u]_{1+\delta/2,2+\delta}$ on the right becomes less than $1/2$. The theorem is proved.

EXERCISE 8.6.2. Prove that there is *no* finite constant $N = N(d,\delta)$ such that for any $u \in C_0^\infty(\mathbb{R}^{d+1})$ we have $|u|_0 \leq N|\Delta u - u_t|_{\delta/2,\delta}$.

8.7. Solvability of the heat equation in the Hölder spaces

Theorem 8.6.1 and Theorem 8.1.7 would provide us with an estimate of the norm $|u|_{1+\delta/2,2+\delta}$ through $|\Delta u - u - u_t|_{\delta/2,\delta}$ if we knew that the norms $|u|_{\delta/2,\delta}$, $|u_t|_0$, $|u_x|_0$ and $|u_{xx}|_0$ are controlled by $|u|_0$ and $[u]_{1+\delta/2,2+\delta}$. This was the way we adopted for elliptic equations. We show that for parabolic equations we could do the same in the following section by proving parabolic interpolation inequalities. In the present section we show how to get an estimate of the entire norm $|u|_{1+\delta/2,2+\delta}$ by using a different idea.

We start with an extension of Theorem 8.6.1 to functions from the space $C^{1+\delta/2,2+\delta}(\mathbb{R}^{d+1})$ rather than $C_0^\infty(\mathbb{R}^{d+1})$. As always we assume that $\delta \in (0,1)$.

LEMMA 8.7.1. *Let $u \in C^{1+\delta/2,2+\delta}(\mathbb{R}^{d+1})$. Define $f = \Delta u - u_t$. Then there exists a constant $N = N(d,\delta)$ such that*

$$[u]_{1+\delta/2,2+\delta} \leq N[f]_{\delta/2,\delta}. \tag{8.7.1}$$

Proof. First let u be infinitely differentiable. Take a function $\xi \in C_0^\infty(\mathbb{R}^{d+1})$ such that $|\xi| \leq 1, \xi(0) = 1$ and for $R \geq 1$ define $\xi_R(z) = \xi(z/R)$, $u_R(z) = u(z)\xi(z/R)$. Then

$$[u_R]_{1+\delta/2,2+\delta} \leq N[\Delta u_R - u_{Rt}]_{\delta/2,\delta}. \tag{8.7.2}$$

Observe that

$$\Delta u_R(z) - u_{Rt}(z) = \xi_R(z)f(z) + R^{-2}u(z)(\Delta\xi)(z/R)$$

$$-R^{-1}u(z)\xi_t(z/R) + R^{-1}2u_{x^i}(z)\xi_{x^i}(z/R), \quad [\xi_R]_{\delta/2,\delta} \leq NR^{-\delta/2}.$$

It is seen that

$$[\Delta u_R - u_{Rt}]_{\delta/2,\delta} \leq [f]_{\delta/2,\delta} + MR^{-\delta/2},$$

where the constant M is independent of R. Using this and Exercise 8.5.6, we get (8.7.1) by sending $R \to \infty$ in (8.7.2).

Now pass to the general case $u \in C^{1+\delta/2,2+\delta}(\mathbb{R}^{d+1})$. Define the functions ζ and u^ε as in Exercise 8.5.7. Obviously u^ε are infinitely differentiable. Therefore (see Exercise 8.5.7),

$$[u]_{1+\delta/2,2+\delta} = \lim_{\varepsilon\downarrow 0}[u^\varepsilon]_{1+\delta/2,2+\delta} \le N \limsup_{\varepsilon\downarrow 0}[\Delta u^\varepsilon - u^\varepsilon_t]_{\delta/2,\delta}$$

$$= N \limsup_{\varepsilon\downarrow 0}[(\Delta u - u_t)^\varepsilon]_{\delta/2,\delta} \le N[\Delta u - u_t]_{\delta/2,\delta}.$$

The lemma is proved.

Now we are ready to estimate $|u|_{1+\delta/2,2+\delta}$.

THEOREM 8.7.2. *Let $u \in C^{1+\delta/2,2+\delta}(\mathbb{R}^{d+1})$, and define $f = \Delta u - u_t - u$. Then for a constant $N = N(d,\delta)$ we have*

$$|u|_{1+\delta/2,2+\delta} \le N|f|_{\delta/2,\delta}. \tag{8.7.3}$$

Proof. Introduce a new independent variable $x^{d+1} \in R$ and define $x' = (x, x^{d+1}) = (x^1, ..., x^{d+1})$, $\phi(x') = \exp(ix^{d+1})$, $v(t,x') = u(t,x)\phi(x')$. Observe that

$$(\Delta_{x'} v - v_t)(t,x') = f(t,x)\phi(x') =: g(t,x').$$

Hence by Lemma 8.7.1

$$[v]_{1+\delta/2,2+\delta;\mathbb{R}^{d+2}} \le N[g]_{\delta/2,\delta;\mathbb{R}^{d+2}} \le N([f]_{\delta/2,\delta}|\phi|_{0;\mathbb{R}^{d+2}} + |f|_0[\phi]_{\delta/2,\delta;\mathbb{R}^{d+2}}). \tag{8.7.4}$$

Further, let D be an operator of differentiation of any order with respect to (t,x) and $D_{(d+1)}$ an operator of differentiation of any order with respect to x^{d+1}. Then

$$\frac{|DD_{(d+1)}v(t,x,x_1^{d+1}) - DD_{(d+1)}v(t,x,x_2^{d+1})|}{|x_1^{d+1} - x_2^{d+1}|^\delta}$$

$$= |Du(t,x)|\frac{|D_{(d+1)}\phi(x_1^{d+1}) - D_{(d+1)}\phi(x_2^{d+1})|}{|x_1^{d+1} - x_2^{d+1}|^\delta},$$

$$\frac{|Dv(z_1,x^{d+1}) - Dv(z_2,x^{d+1})|}{\rho^\delta(z_1,z_2)} = |\phi(x^{d+1})|\frac{|Du(z_1) - Du(z_2)|}{\rho^\delta(z_1,z_2)}.$$

Taking various combinations of D and $D_{(d+1)}$, one easily sees that $|u|_{1+\delta/2,2+\delta}$ is less than an absolute constant times the left–hand side of (8.7.4). This proves the theorem.

THEOREM 8.7.3. *For any $f \in C^{\delta/2,\delta}(\mathbb{R}^{d+1})$ there exists a unique function $u \in C^{1+\delta/2,2+\delta}(\mathbb{R}^{d+1})$ satisfying the equation*

$$\Delta u - u_t - u = f \quad in \quad \mathbb{R}^{d+1}.$$

Proof. Uniqueness we know from Theorem 8.1.7. To prove existence notice that we have it by Theorem 8.4.2 if $f \in C_0^\infty(\mathbb{R}^{d+1})$. By using the cut–off function ζ from the proof of Theorem 8.7.2 and convolutions, for any given $f \in C^{\delta/2,\delta}(\mathbb{R}^{d+1})$ we can easily find a sequence of functions $f_n \in C_0^\infty(\mathbb{R}^{d+1})$ such that $f_n \to f$ at any point and the $C^{\delta/2,\delta}(\mathbb{R}^{d+1})$–norms of f_n are uniformly bounded. Take such a sequence and denote by u_n the corresponding solutions. By Theorem 8.7.2 the $C^{1+\delta/2,2+\delta}(\mathbb{R}^{d+1})$–norms of u_n are bounded. It follows that there is a subsequence $u_{n(k)}$ which converges to a function u at any point along with the derivatives with respect to $t, x^i x^j$. Furthermore, $u \in C^{1+\delta/2,2+\delta}(\mathbb{R}^{d+1})$, and it remains only to let $k \to \infty$ in $\Delta u_{n(k)} - u_{n(k)t} - u_{n(k)} = f_{n(k)}$. The theorem is proved.

EXERCISE* 8.7.4. Let $u \in C^{1+\delta/2,2+\delta}(\mathbb{R}^{d+1})$, $\lambda > 0$ and define $f = \Delta u - u_t - \lambda u$. Prove that for the constant N from Theorem 8.7.2 we have

$$[u]_{1+\delta/2,2+\delta} + \lambda^{\delta/2}(|u_{xx}|_0 + |u_t|_0) + \lambda^{(1+\delta)/2}|u_x|_0 + \lambda^{1+\delta/2}|u|_0$$

$$\leq N([f]_{\delta/2,\delta} + \lambda^{\delta/2}|f|_0).$$

EXERCISE* 8.7.5. Prove that Theorem 8.7.3 is true for the equation $\Delta u - u_t - \lambda u = f$ in $\mathbb{R}^{d+1}$ if $\lambda > 0$.

8.8. Parabolic interpolation inequalities

In order to be able to treat parabolic equations with variable coefficients we need to prove certain interpolation inequalities. Let Ω be an open convex cone in $\mathbb{R}^d$, $T \in (-\infty, \infty]$ and $Q = (-\infty, T) \times \Omega$ or $Q = (T, \infty) \times \Omega$. Remember the notation $\theta(\Omega)$ introduced in Sec. 3.2, and remember that $\delta \in (0,1)$, although actually our results here are valid for $\delta = 1$ as well.

THEOREM 8.8.1. *There exists a constant $N = N(d, \theta(\Omega))$ such that for any $\varepsilon > 0$ and any $u \in C^{1+\delta/2,2+\delta}(Q)$ we have*

$$|u_t|_{0;Q} \leq \varepsilon[u_t]_{\delta/2,\delta;Q} + N\varepsilon^{-2/\delta}|u|_{0;Q}. \tag{8.8.1}$$

$$|u_{xx}|_{0;Q} \leq \varepsilon[u]_{1+\delta/2,2+\delta;Q} + N\varepsilon^{-2/\delta}|u|_{0;Q}. \tag{8.8.2}$$

$$|u_x|_{0;Q} \leq \varepsilon[u]_{1+\delta/2,2+\delta;Q} + N\varepsilon^{-1/(1+\delta)}|u|_{0;Q}. \tag{8.8.3}$$

$$[u_x]_{\delta/2,\delta;Q} \leq \varepsilon[u]_{1+\delta/2,2+\delta;Q} + N\varepsilon^{-(1+\delta)}|u|_{0;Q}. \tag{8.8.4}$$

$$[u]_{\delta/2,\delta;Q} \leq \varepsilon[u]_{1+\delta/2,2+\delta;Q} + N\varepsilon^{-\delta/2}|u|_{0;Q}. \tag{8.8.5}$$

Proof. Parabolic dilations, that is, transformations of the form $u(t,x) \to u(R^2t, Rx)$, show at once that we need to consider only the case $\varepsilon = 1$ and prove (8.8.1) through (8.8.5) with $N[u]_{1+\delta/2,2+\delta;Q}$ instead of $\varepsilon[u]_{1+\delta/2,2+\delta;Q}$.

Take $z = (t,x) \in Q$ and notice that for a $\theta \in (0,1)$ we have

$$|u_t(z)| \leq |u_t(z) - [u(t+1,x) - u(t,x)]| + 2|u|_{0;Q}$$

$$= |u_t(t,x) - u_t(t+\theta,x)| + 2|u|_{0;Q} \leq 2|u|_{0;Q} + [u_t]_{\delta/2,\delta;Q}.$$

This proves (8.8.1).

Actually we have just repeated a part of the proof of Theorem 3.2.1 for $d = 1$. We can also fix t and apply this theorem with respect to x. This obviously yields (8.8.2) and (8.8.3). Owing to the same theorem and Exercise 8.5.5, we get (8.8.5) by fixing separately t and x.

In the remaining assertion (8.8.4) the estimate of the Hölder constant of $u_x$with respect to x again follows from Theorem 3.2.1. Therefore, it remains to estimate $|u_{x^i}(t,x) - u_{x^i}(s,x)|$.

By the above for any $(t,x),(t,y) \in Q$

$$|u_{x^i}(t,x) - u_{x^i}(s,x)| \le |u_{x^i}(t,x) - u_{x^i}(t,y)| + |u_{x^i}(s,x) - u_{x^i}(s,y)|$$

$$+ |u_{x^i}(t,y) - u_{x^i}(s,y)| \le |u_{x^i}(t,y) - u_{x^i}(s,y)|$$

$$+ N|x-y|^\delta(|u_x|_{0;Q} + |u_{xx}|_{0;Q}).$$

This shows that to finish the proof of (8.8.4) we have to estimate $|u_{x^i}(t,y) - u_{x^i}(s,y)|$ through $|t-s|^{\delta/2}$ times the sum of $|u|_{0;Q}$ and $[u]_{1+\delta/2,2+\delta;Q}$ only for y at distance $\theta(\Omega)|t-s|^{1/2}$ from x and, hence, at a distance at least $|t-s|^{1/2}$ from ∂Q. In addition, obviously we may consider only $|t-s| \le 1$. For $\varepsilon = |t-s|^{1/2}$ and certain $\theta_i \in (0,1)$ we have

$$|u_{x^i}(t,y) - u_{x^i}(s,y)| \le |u_{x^i}(t,y) - \varepsilon^{-1}[u(t,y+\varepsilon e_i) - u(t,y)]|$$

$$+ |u_{x^i}(s,y) - \varepsilon^{-1}[u(s,y+\varepsilon e_i) - u(s,y)]| + \varepsilon^{-1}|u(t,y+\varepsilon e_i) - u(s,y+\varepsilon e_i)|$$

$$+ \varepsilon^{-1}|u(t,y) - u(s,y)| \le |u_{x^i}(t,y) - u_{x^i}(t,y+\theta_1\varepsilon e_i)|$$

$$+ |u_{x^i}(s,y) - u_{x^i}(s,y+\theta_2\varepsilon e_i)| + 2\varepsilon|u_t|_{0;Q} \le N\varepsilon(|u_{x^ix^i}|_{0;Q} + |u_t|_{0;Q}).$$

Since $\varepsilon \le |t-s|^{\delta/2}$ and we have already estimated $|u_{x^ix^i}|_{0;Q}$ and $|u_t|_{0;Q}$ in (8.8.2) and (8.8.1), estimate (8.8.4) and the theorem are completely proved.

In the following exercises by Q we mean the domain from Theorem 8.8.1.

EXERCISE* 8.8.2. Prove the following *multiplicative inequalities*

$$|u_t|_{0;Q} + |u_{xx}|_{0;Q} \le NU_{2+\delta}^{2/(2+\delta)}U_0^{1-2/(2+\delta)},$$

$$|u_x|_{0;Q} \le NU_{2+\delta}^{1/(2+\delta)}U_0^{1-1/(2+\delta)},$$

$$[u_x]_{\delta/2,\delta;Q} \le NU_{2+\delta}^{(1+\delta)/(2+\delta)}U_0^{1-(1+\delta)/(2+\delta)},$$

$$[u]_{\delta/2,\delta;Q} \le NU_{2+\delta}^{\delta/(2+\delta)}U_0^{1-\delta/(2+\delta)},$$

where $N = N(d)$, $U_0 = |u|_{0;Q}$, $U_{2+\delta} = [u]_{1+\delta/2,2+\delta;Q}$. It is easy to memorize these inequalities if one takes into account that the inequalities are invariant under multiplication of u by a constant and also under parabolic dilations.

EXERCISE 8.8.3. The proof of Theorem 8.7.2 given above hinges on a trick of introducing a new independent variable. Prove this theorem on the basis of Lemma 8.7.1 and Theorem 8.1.7 and interpolation inequalities after noticing that, for instance, $[f]_{\delta/2,\delta} \leq [f-u]_{\delta/2,\delta} + [u]_{\delta/2,\delta}$.

EXERCISE 8.8.4 (first step toward quasilinear equations). Prove that if $F(z,\xi)$ is a real–valued function of $z \in \mathbb{R}^{d+1}, \xi \in \mathbb{R}^d$, and for some constants $n, k \geq 0$ we have $[F(\cdot,\xi)]_{\delta/2,\delta} \leq |\xi|^n$ and $|F_\xi| \leq |\xi|^k$, then for any $u \in C^{1+\delta/2,2+\delta}(\mathbb{R}^{d+1})$ in notation from Exercise 8.8.2 it holds that

$$[F(\cdot,u_x(\cdot))]_{\delta/2,\delta} \leq NU_{2+\delta}^{n/(2+\delta)}U_0^{n-n/(2+\delta)} + NU_{2+\delta}^{(k+1+\delta)/(2+\delta)}U_0^{k+1-(k+1+\delta)/(2+\delta)}.$$

EXERCISE 8.8.5 (cf. Exercise 3.2.8). Let $F(z,\xi)$ be a function of $z \in \mathbb{R}^{d+1}$ and $\xi \in \mathbb{R}^{d_1}$. Assume that for some constants $\delta_1, \delta_2 \in (0,1]$ we have

$$|F(z_1,\xi) - F(z_2,\xi)| \leq \rho^{\delta_1\delta_2}(z_1,z_2), \quad |F(z_1,\xi) - F(z_1,\eta)| \leq |\xi-\eta|^{\delta_2}$$

for any $z_1, z_2 \in \mathbb{R}^d$ and $\xi, \eta \in \mathbb{R}^{d_1}$. For simplicity of formulation also assume that $\delta_1\delta_2 < 1$. Prove that for any $\mathbb{R}^{d_1}$-valued function $u \in C^{\delta_1/2,\delta_1}(\mathbb{R}^{d+1})$ we have $F(\cdot,u(\cdot)) \in C^{\delta_1\delta_2/2,\delta_1\delta_2}(\mathbb{R}^{d+1})$. Surprisingly enough, generally speaking, the operator $F: u \to F(\cdot,u(\cdot))$ is *not* continuous as an operator from $C^{\delta_1/2,\delta_1}(\mathbb{R}^{d+1})$ into $C^{\delta_1\delta_2/2,\delta_1\delta_2}(\mathbb{R}^{d+1})$ if $\delta_1, \delta_2 \in (0,1)$.

EXERCISE 8.8.6. Prove that for $u \in C^{1+\delta/2,2+\delta}(Q)$ and $x \in \Omega$ and all t, s we have

$$|u_{x^i}(t,x) - u_{x^i}(s,x)| \leq N(d,\delta)|t-s|^{(1+\delta)/2}\{|u|_{0;Q} + [u]_{1+\delta/2,2+\delta;Q}\}.$$

REMARK 8.8.7. Above we have considered the parabolic Hölder spaces of functions having either zero or two derivatives with respect to x. The question arises about defining naturally a space $C^{(1+\delta)/2,1+\delta}(\mathbb{R}^{d+1})$ of functions having only one derivative in x. The above results show that with respect to the parabolic metric, one derivative in t is worth two derivatives in x. This suggests that $C^{(1+\delta)/2,1+\delta}(\mathbb{R}^{d+1})$ should be defined as the space of all functions with finite norm

$$|u|_0 + |u_x|_{\delta/2,\delta} + \sup_{s\neq t,x} \frac{|u(t,x) - u(s,x)|}{|t-s|^{(1+\delta)/2}}.$$

By the way, the above exercise shows that under this definition we have the following natural property: the operators D_i are bounded operators from $C^{1+\delta/2,2+\delta}(\mathbb{R}^{d+1})$ into $C^{(1+\delta)/2,1+\delta}(\mathbb{R}^{d+1})$.

EXERCISE 8.8.8. By using Exercise 3.3.7 prove that for $u \in C^{1+\delta/2,2+\delta}(\mathbb{R}^{d+1})$ and $\delta \in (0,1]$ one has $[u]_{1/2,1} \leq N(d,\delta)[u]_{1+\delta/2,2+\delta}^{(1-\delta)/2}[u]_{\delta/2,\delta}^{(1+\delta)/2}$.

EXERCISE 8.8.9 (sharp parabolic embedding theorem). On the basis of Exercises 8.8.8 and 8.5.10 and Theorem 8.6.1 prove that if $1 + d/2 < p < 2 + d$, then for $\delta := 2 - (d+2)/p$ and any $u \in C_0^\infty(\mathbb{R}^d)$ we have $\delta \in (0,1)$ and

$$[u]_{\delta/2,\delta} \leq N(d,p)||\Delta u - u_t||_{L_p(\mathbb{R}^{d+1})}.$$

8.9. The Schauder a priori estimates

Take the operator L from (8.0.2) and assume that, for any t, it is a second–order elliptic operator with real coefficients and a constant of ellipticity $\kappa > 0$. Also let $|a, b, c|_{\delta/2,\delta} \leq K$, where K is a fixed constant and $\delta \in (0, 1)$.

First, we extend Theorem 8.7.2 to the arbitrary elliptic operator of *main type* with *constant* coefficients. Its proof is based on a simple change of coordinates.

LEMMA 8.9.1. *Assume that the coefficients a^{ij} are constant and $b^i = c = 0$. Then there exists a constant N depending only on d, κ, δ, K such that for any $u \in C^{1+\delta/2,2+\delta}(\mathbb{R}^{d+1})$ we have*

$$|u|_{1+\delta/2,2+\delta} \leq N|Lu - u_t - u|_{\delta/2,\delta}. \tag{8.9.1}$$

To prove this lemma it suffices to take a nondegenerate matrix A such that $AaA^T = I$, where I is the $d \times d$ identity matrix, then define $v(t, Ax) = u(t, x)$, $g(t, Ax) = f(t, x)$ and observe the simple relationship between norms of u and v and that $g = \Delta v - v_t - v$ (cf. Remark 2.8.1).

Now we are ready to prove (8.9.1) for operators with variable coefficients.

THEOREM 8.9.2. *There exists a constant N depending only on d, κ, δ, K such that estimate (8.9.1) holds for any $u \in C^{1+\delta/2,2+\delta}(\mathbb{R}^{d+1})$.*

Proof. Denote $f = Lu - u_t - u$. Owing to Theorem 8.8.1, we need to estimate only $[u]_{1+\delta/2,2+\delta}$ and $|u|_0$ through $|f|_{\delta/2,\delta}$. Also, by Theorem 8.1.7 we have $|u|_0 \leq |f|_0$. Therefore, we need to estimate only $[u]_{1+\delta/2,2+\delta}$. To simplify the presentation we first assume that $b^i \equiv c \equiv 0$.

Fix a small constant $\gamma > 0$, which will be specified later, and find two points z_1, z_2 such that

$$\frac{|u_t(z_1) - u_t(z_2)|}{\rho^\delta(z_1, z_2)} \geq \frac{1}{2}\,[u_t]_{\delta/2,\delta}.$$

If $\rho(z_1, z_2) \geq \gamma$, then the fraction on the left is less than $\gamma^{-\delta}2|u_t|_0$, so that by the interpolation inequalities

$$[u_t]_{\delta/2,\delta} \leq \frac{1}{4}\,[u]_{1+\delta/2,2+\delta} + N(\gamma)|u|_0 \leq \frac{1}{4}\,[u]_{1+\delta/2,2+\delta} + N(\gamma)|f|_0, \tag{8.9.2}$$

where, and hereafter, by $N(\gamma)$ we denote various constants depending only on γ, d, κ, δ, and K.

In the second case: $\rho(z_1, z_2) \leq \gamma$, take $\zeta \in C_0^\infty(\mathbb{R}^{d+1})$ so that $\zeta(z) = 1$ if $\rho(z, 0) \leq 1$ and $\zeta(z) = 0$ if $\rho(z, 0) \geq 2$ and define

$$\xi(t, x) = \zeta((t - t_1)/\gamma^2, (x - x_1)/\gamma).$$

Now we "freeze" the coefficients of L. Observe that by Lemma 8.9.1 ($b = c = 0$)

$$[u_t]_{\delta/2,\delta} \leq 2\frac{|u_t(z_1) - u_t(z_2)|}{\rho^\delta(z_1, z_2)} \leq [(u\xi)]_{1+\delta/2,2+\delta} \leq N|L(z_1)(u\xi) - (u\xi)_t - u\xi|_{\delta/2,\delta}$$

$$\leq N|L(u\xi) - (u\xi)_t - u\xi|_{\delta/2,\delta} + N|(L(z_1) - L)(u\xi)|_{\delta/2,\delta}.$$

Since $\xi(z) = 0$ for $\rho(z, z_1) \geq 2\gamma$, the last norm can be taken over $\{z : \rho(z, z_1) < 2\gamma\}$ (where $|a(z) - a(z_1)| \leq N\gamma^\delta$), and it is seen that

$$|(L(z_1)-L)(u\xi)|_{\delta/2,\delta}\leq N\gamma^{\delta}[(u\xi)_{xx}]_{\delta/2,\delta}+N|(u\xi)_{xx}|_0$$

$$\leq N\gamma^{\delta}[u]_{1+\delta/2,2+\delta}+N(\gamma)|u|_0\leq N\gamma^{\delta}[u]_{1+\delta/2,2+\delta}+N(\gamma)|f|_0.$$

Also

$$|L(u\xi)-(u\xi)_t-u\xi|_{\delta/2,\delta}=|f\xi+uL\xi+2a^{ij}u_{x^i}\xi_{x^j}|_{\delta/2,\delta}$$

$$\leq N(\gamma)\Big(|f|_{\delta/2,\delta}+|u|_{\delta/2,\delta}+|u_x|_{\delta/2,\delta}\Big)$$

$$\leq\gamma^{\delta}[u]_{1+\delta/2,2+\delta}+N(\gamma)|f|_{\delta/2,\delta}.$$

Hence in this case

$$[u_t]_{\delta/2,\delta}\leq N\gamma^{\delta}[u]_{1+\delta/2,2+\delta}+N(\gamma)|f|_{\delta/2,\delta}.$$

Comparing this with (8.9.2), we see that always

$$[u_t]_{\delta/2,\delta}\leq(N\gamma^{\delta}+1/4)[u]_{1+\delta/2,2+\delta}+N(\gamma)|f|_{\delta/2,\delta}.$$

Obviously, we can proceed in the same way while estimating $[u_{x^ix^j}]_{\delta/2,\delta}$, fixing i,j and taking (new) z_1,z_2 such that

$$\frac{|u_{x^ix^j}(z_1)-u_{x^ix^j}(z_2)|}{\rho^{\delta}(z_1,z_2)}\geq\frac{1}{2}[u_{x^ix^j}]_{\delta/2,\delta}.$$

Then we find

$$[u_{x^ix^j}]_{\delta/2,\delta}\leq(N\gamma^{\delta}+1/(4d^2))[u]_{1+\delta/2,2+\delta}+N(\gamma)|f|_{\delta/2,\delta}.$$

Summing up these estimates, we conclude

$$[u]_{1+\delta/2,2+\delta}\leq(N_1\gamma^{\delta}+1/2)[u]_{1+\delta/2,2+\delta}+N(\gamma)|f|_{\delta/2,\delta},$$

and to get (8.9.1) it remains only to choose γ so that $N_1\gamma^{\delta}\leq 1/4$.

Now we have to deal with operators containing lower–order terms. From the previous result we have

$$[u]_{1+\delta/2,2+\delta}\leq N(|f|_{\delta/2,\delta}+|b^iu_{x^i}|_{\delta/2,\delta}+|cu|_{\delta/2,\delta}).$$

As always the additional terms on the right can be estimated through $\gamma[u]_{1+\delta/2,2+\delta}$ plus $N(\gamma)|u|_0$; and then by choosing small γ, collecting like terms and using again that $|u|_0\leq|f|_0$, we arrive at (8.9.1). The theorem is proved.

EXERCISE* 8.9.3. Prove that there exists a constant N depending only on d,κ,δ,K such that for any $\lambda\geq 1$ and $u\in C^{1+\delta/2,2+\delta}(\mathbb{R}^{d+1})$ we have

$$|u|_{1+\delta/2,2+\delta}\leq N([f]_{\delta/2,\delta}+\lambda^{\delta/2}|f|_0),\quad |u|_0\leq\lambda^{-1}|f|_0,$$

where $f=Lu-u_t-\lambda u$.

EXERCISE* 8.9.4. By using Exercises 8.9.3 and 8.8.2 prove that there exists a constant N depending only on d, κ, δ, K such that for any $\lambda \geq 1$ and $u \in C^{1+\delta/2,2+\delta}(\mathbb{R}^{d+1})$ we have

$$|u|_{\delta/2,\delta} + |u_x|_{\delta/2,\delta} \leq N\lambda^{(\delta-1)/2}|Lu - u_t - \lambda u|_{\delta/2,\delta}.$$

EXERCISE 8.9.5. Let $F(\xi)$ be a continuously differentiable real–valued function on $\mathbb{R}^d$ such that $F(0) = 0$ and $|F_\xi(\xi)| \leq |\xi|^k$ with a constant $k \in [0, 1)$. Prove that if $u \in C^{1+\delta/2,2+\delta}(\mathbb{R}^{d+1})$ and $f \in C^{\delta/2,\delta}(\mathbb{R}^{d+1})$ and $\Delta u + F(\operatorname{grad} u) - u_t - u = f$, then $|u|_{1+\delta/2,2+\delta} \leq N(d, k, \delta, |f|_{\delta/2,\delta})$.

8.10. The existence theorems

Let L be an operator as in Sec. 8.9 and $\delta \in (0, 1)$.

THEOREM 8.10.1. *Assume that $c \leq -\lambda$ for a constant $\lambda > 0$. Also let $T \in (-\infty, \infty]$ (including ∞) and define $Q = (-\infty, T) \times \mathbb{R}^d$. Then for any $f \in C^{\delta/2,\delta}(Q)$ there exists a unique solution $u \in C^{1+\delta/2,2+\delta}(Q)$ of equation (8.0.1) in Q. In addition, there exists a constant N depending only on $d, \kappa, \delta, K, \lambda$ such that*

$$|u|_{1+\delta/2,2+\delta;Q} \leq N|f|_{\delta/2,\delta;Q} = N|Lu - u_t|_{\delta/2,\delta;Q}. \tag{8.10.1}$$

Proof. It suffices to prove the theorem for $T = \infty$. Indeed, if T is finite, continue f for $t \geq T$ by the formula $f(t, x) = f(T, x)$. This will not affect the $C^{\delta/2,\delta}$–norm of f, and if the theorem is true for $T = \infty$, we will get a function u for which (8.10.1) holds even with $\mathbb{R}^{d+1}$ instead of Q on the left. Finally, by Theorem 8.1.7 any other function given for $t \leq T$ and satisfying (8.0.1) in Q coincides in Q with the one found above in $\mathbb{R}^{d+1}$.

So let $T = \infty$. On the basis of Theorems 8.7.3 and 8.9.2 on a priori estimates the proof of this theorem is made by the method of continuity exactly as in Sec. 4.4 if the coefficient c can be represented as $c' - 1$ with $c' \leq 0$, or in other words, if $c \leq -1$. The general case is reduced to this one by changing the unknown function $v(\lambda t, x) = u(t, x)$ (and dividing through the equation by λ). The theorem is proved.

If the coefficients of L and f are independent of t, then the function $u(t+\gamma, x)$ satisfies the same equation as u for any number γ, and it follows from uniqueness that u is independent of t. In this way we arrive at the following version of Theorem 4.3.2.

COROLLARY 8.10.2. *Assume that the coefficients of L and f are independent of t and $c \leq -\lambda$. Also let $f \in C^\delta(\mathbb{R}^d)$. Then there exists a unique solution $u \in C^{2+\delta}(\mathbb{R}^d)$ of the equation $Lu = f$ in $\mathbb{R}^d$. In addition, with a constant N depending only on $\lambda, d, \delta, \kappa, K$ we have $|u|_{2+\delta} \leq N|f|_\delta$.*

REMARK 8.10.3. By Theorem 8.10.1 for any function f of class $C^{\delta/2,\delta}(\mathbb{R}^{d+1})$ and any $\lambda > 0$ there exists a unique solution $u \in C^{1+\delta/2,2+\delta}(\mathbb{R}^{d+1})$ of the equation $Lu - u_t - \lambda u + f = 0$ in $\mathbb{R}^{d+1}$. We call this solution $\mathcal{R}_\lambda f$.

By Exercises 8.9.3 and 8.9.4 for $\lambda \geq 1$ we have

$$|\mathcal{R}_\lambda f|_{1+\delta/2,2+\delta} \leq N\Big([f]_{\delta/2,\delta} + \lambda^{\delta/2}|f|_0\Big), \quad |\mathcal{R}_\lambda f|_0 \leq \lambda^{-1}|f|_0,$$

$$|\mathcal{R}_\lambda f|_{\delta/2,\delta} + |(\mathcal{R}_\lambda f)_x|_{\delta/2,\delta} \leq N\lambda^{(\delta-1)/2}|f|_{\delta/2,\delta},$$

the constant N depending only on d, κ, δ, K.

EXERCISE 8.10.4. Let $F(\xi)$ be a continuously differentiable function given on $\mathbb{R}^d$ and having bounded derivatives. Prove that for any $f \in C^{\delta/2,\delta}(\mathbb{R}^{d+1})$ there exists a unique solution $u \in C^{1+\delta/2,2+\delta}(\mathbb{R}^{d+1})$ of the equation $\Delta u + F(\operatorname{grad} u) - u_t - u = f$.

EXERCISE 8.10.5. Given a constant $k \in [0,1)$ and $f \in C^{\delta/2,\delta}(\mathbb{R}^{d+1})$, prove that there exists a unique solution $u \in C^{1+\delta/2,2+\delta}(\mathbb{R}^{d+1})$ of the equation $\Delta u + |\operatorname{grad} u|^{k+1} - u_t - u = f$.

8.11. Interior a priori estimates

We take an operator L as in Sec. 8.9 and $\delta \in (0,1)$.

THEOREM 8.11.1. *For any $R > 0$ there is a constant N depending only on R, κ, δ, K, d such that if $u \in C^{1+\delta/2,2+\delta}(Q_{3R})$, then*

$$|u|_{1+\delta/2,2+\delta;Q_R} \le N(|Lu - u_t|_{\delta/2,\delta;Q_{2R}} + |u|_{0;Q_{2R}}). \tag{8.11.1}$$

Proof. Parabolic dilations show that without loss of generality we can assume $R = 1$. Let $R_n = \sum_{j=0}^{n} 2^{-j}$, $n = 0, 1, 2, \ldots$. As in the proof of Theorem 7.1.1 we need some functions $\zeta_n \in C_0^\infty(\mathbb{R}^{d+1})$, $\zeta_n(x) = 1$ in $C_n := (-R_n, 0) \times B_{R_n}$, $\zeta_n(x) = 0$ outside $(-R_{n+1}, R_{n+1}) \times B_{R_{n+1}}$ and

$$|\zeta_n, \zeta_{nx}, \zeta_{nt}, \zeta_{nxx}|_{\delta/2,\delta} \le N 2^{3n} = N\rho^{-n},$$

where $\rho = 2^{-3} < 1$ and $N = N(d, k, m)$. To construct them it suffices to take $h \in C_b^\infty(R)$ such that $h(t) = 1$ for $t \le 1$, $h(t) = 0$ for $t \ge 2$ and $0 \le h \le 1$ and define

$$\xi_n(x) = h(2^{n+1}(|x| - R_n + 2^{-(n+1)})),$$

$$\eta_n(t) = h(2^{n+1}(|t| - R_n + 2^{-(n+1)})), \quad \zeta_n = \xi_n \eta_n.$$

Finally, let $f := Lu - u_t$.

Put $u\zeta_n$ in (8.10.1) with $Q = (-\infty, 0) \times \mathbb{R}^d$ to get

$$\alpha_n := |u\zeta_n|_{1+\delta/2,2+\delta;Q} \le N|L(u\zeta_n) - (u\zeta_n)_t - u\zeta_n|_{\delta/2,\delta;Q_2}$$

$$\le N\Big(|L(u\zeta_n) - (u\zeta_n)_t|_{\delta/2,\delta;Q} + |u\zeta_n|_{\delta/2,\delta;Q}\Big) \le N|Lu - u_t|_{\delta/2,\delta;Q_2}$$

$$+ N\rho^{-n}\Big(|u_x|_{\delta/2,\delta;C_{n+1}} + |u|_{\delta/2,\delta;C_{n+1}}\Big). \tag{8.11.2}$$

Further, take $\zeta \in C_0^\infty(\mathbb{R}^{d+1})$ such that $\zeta = 1$ in Q_2 and $\zeta = 0$ outside $Q_3 \cup (-Q_3)$. Owing to the assumption $u \in C^{1+\delta/2,2+\delta}(Q_3)$ and Theorem 8.8.1, we have

$$|u_x|_{\delta/2,\delta;C_{n+1}} + |u|_{\delta/2,\delta;C_{n+1}} \le |(\zeta u)_x|_{\delta/2,\delta;Q} + |\zeta u|_{\delta/2,\delta;Q} =: M < \infty.$$

Therefore, it follows from (8.11.2) that α_n does not increase too fast: $\alpha_n \le N(1+M)\rho^{-n}$, and

$$\alpha_n \le N|f|_{\delta/2,\delta;Q_2} + N\rho^{-n}\Big(|(u\zeta_{n+1})_x|_{\delta/2,\delta;Q} + |u\zeta_{n+1}|_{\delta/2,\delta;Q}\Big). \tag{8.11.3}$$

By interpolation inequalities (see Theorem 8.8.1 with $Q = (-\infty, 0) \times \mathbb{R}^d$) there exist constants $N, m > 0$ depending only on δ, d, such that for $\varepsilon \in (0, 1]$ we have

$$|(u\zeta_{n+1})_x|_{\delta/2,\delta;Q} + |u\zeta_{n+1}|_{\delta/2,\delta;Q}$$

$$\le \varepsilon|u\zeta_{n+1}|_{1+\delta/2,2+\delta;Q} + N\varepsilon^{-m}|u\zeta_{n+1}|_{0;Q} \le \varepsilon\alpha_{n+1} + N\varepsilon^{-m}|u|_{0;Q_2}.$$

This and (8.11.3) imply that

$$\alpha_n \le \varepsilon\rho^{-n}\alpha_{n+1} + N\varepsilon^{-m}\rho^{-n}|u|_{0;Q_2} + N|f|_{\delta/2,\delta;Q_2},$$

and since $\varepsilon \in (0, 1]$ is arbitrary,

$$\alpha_n \le \varepsilon\alpha_{n+1} + N\varepsilon^{-m}\rho^{-n(m+1)}|u|_{0;Q_2} + N|f|_{\delta/2,\delta;Q_2},$$

$$\varepsilon^n\alpha_n \le \varepsilon^{n+1}\alpha_{n+1} + N\varepsilon^{-m}\varepsilon^n\rho^{-n(m+1)}|u|_{0;Q_2} + N\varepsilon^n|f|_{\delta/2,\delta;Q_2}$$

for any $\varepsilon \in (0, 1]$ too. Now take and fix ε so that $\varepsilon\rho^{-m-1} \le 1/2$. Then $N\varepsilon^{-m} = N$ and

$$\varepsilon^n\alpha_n \le \varepsilon^{n+1}\alpha_{n+1} + N2^{-n}|u|_{0;Q_2} + N\varepsilon^n|f|_{\delta/2,\delta;Q_2},$$

$$\sum_{n=0}^{\infty}\alpha_n\varepsilon^n \le \sum_{n=1}^{\infty}\alpha_n\varepsilon^n + N|u|_{0;Q_2}\sum_{n=0}^{\infty}2^{-n} + N|f|_{\delta/2,\delta;Q_2}\sum_{n=0}^{\infty}\varepsilon^n.$$

To prove (8.11.1) (with $R = 1$) it remains only to collect like terms in the last inequality after noticing that its left–hand side is finite, since $\varepsilon < 1$ and $\alpha_n\varepsilon^n \le N(1+M)\rho^{-n}\varepsilon^n \le N(1+M)(1/2)^n\rho^{mn}$. The theorem is proved.

REMARK 8.11.2. Let $Q^1 \subset Q^2$ be domains in $\mathbb{R}^{d+1}$. Assume that for a number $R \in (0, 1]$, for any point $z \in Q^1$ we have $z + Q_{3R} \subset Q^2$. Then by considering $z + Q_{3R}$, $z + Q_R$ instead of Q_{3R}, Q_R for all possible z we see that there is a constant N depending only on R, κ, δ, K, d and the diameter of Ω' such that if $u \in C^{1+\delta/2,2+\delta}(Q^2)$, then

$$|u|_{1+\delta/2,2+\delta;Q^1} \le N(|Lu - u_t|_{\delta/2,\delta;Q^2} + |u|_{0;Q^2}).$$

8.12. Better regularity of solutions

The results of Sec. 8.11 allow us to prove that better smoothness of f and a, b, c implies better smoothness of u. As always we assume that $\delta \in (0, 1)$.

THEOREM 8.12.1. *Assume that for an $R > 0$ and a given integer $k \ge 0$ we have $|D^\alpha a|_{\delta/2,\delta;Q_{2R}}$, $|D^\alpha b|_{\delta/2,\delta;Q_{2R}}$, $|D^\alpha c|_{\delta/2,\delta;Q_{2R}} \le K$ for any multi–index $\alpha \in \mathbb{R}^d$ with $|\alpha| \le k$. Assume that $u \in C^{1+\delta/2,2+\delta}(Q_{2R})$ and $D^\alpha(Lu - u_t) \in C^{\delta/2,\delta}(Q_{2R})$ for $|\alpha| \le k$. Then we have $D^\alpha u \in C^{1+\delta/2,2+\delta}(Q_R)$ if $|\alpha| \le k$, and there is a constant N depending only on $R, K, \delta, k, d, \kappa$ such that*

$$\sum_{|\alpha|\le k}|D^\alpha u|_{1+\delta/2,2+\delta;Q_R} \le N\sum_{|\alpha|\le k}|D^\alpha(Lu - u_t)|_{\delta/2,\delta;Q_{2R}} + N|u|_{0;Q_{2R}}. \tag{8.12.1}$$

Proof. If $k=0$, we have nothing to prove due to Theorem 8.11.1 (and Remark 8.11.2). Let $k\geq 1$. Fix $h\in(0,R/4)$, $r=1,...,d$, then denote $f=Lu-u_t$, $v=\delta_{h,r}u$ and observe that

$$a^{ij}(t,x)v_{x^ix^j}(t,x)+b^i(t,x)v_{x^i}(t,x)+c(t,x)v(x)=\delta_{h,r}f(t,x)$$

$$-u_{x^ix^j}(t,x)\delta_{h,r}a^{ij}(t,x+he_r)-u_{x^i}(t,x)\delta_{h,r}b^i(t,x+he_r)-u(t,x)\delta_{h,r}c(t,x+he_r).$$

By our assumptions the $C^{\delta/2,\delta}(Q_{3R/2})$–norm of the right–hand side is bounded by a constant independent of h. It follows by Remark 8.11.2 that the $C^{1+\delta/2,2+\delta}(Q_R)$–norm of v is bounded by a constant independent of h. In turn, this implies that $D_r u\in C^{1+\delta/2,2+\delta}(Q_R)$.

Upon applying these results to cylinders Q_ρ and moving them inside Q_{2R} we see that actually $D_r u\in C^{1+\delta/2,2+\delta}_{loc}(Q_{2R})$. Therefore, we can differentiate the equation $f=Lu-u_t$ in Q_{2R} to obtain that in Q_{2R}

$$Lv=f_{x^r}-u_{x^ix^j}a^{ij}_{x^r}-u_{x^i}b^i_{x^r}-uc_{x^r}, \tag{8.12.2}$$

where $v=u_{x^r}$. By applying Remark 8.11.2 twice we conclude

$$|v|_{1+\delta/2,2+\delta;Q_R}\leq N\Big(|f_{x^r}|_{\delta/2,\delta;Q_{3R/2}}+|u|_{1+\delta/2,2+\delta;Q_{3R/2}}\Big)$$

$$\leq N\Big(|f_{x^r}|_{\delta/2,\delta;Q_{2R}}+|f|_{\delta/2,\delta;Q_{2R}}+|u|_{0;Q_{2R}}\Big).$$

This proves the theorem for $k=1$. Now assume that the theorem is proved for $k=k_0$ and its assumptions are satisfied for $k=k_0+1$. Observe then that the assertion of the theorem is true not only for the cylinders Q_R and Q_{2R} but for any two domains Q^1 and Q^2 satisfying the conditions from Remark 8.11.2. Therefore, by the assumption from (8.12.2) and (8.12.1) we get

$$\sum_{|\alpha|\leq k_0}|D^\alpha v|_{1+\delta/2,2+\delta;Q_R}\leq N\sum_{|\alpha|\leq k_0}|D^\alpha f_{x^r}|_{\delta/2,\delta;Q_{3R/2}}+N\sum_{|\alpha|\leq k_0+2}|D^\alpha u|_{\delta/2,\delta;Q_{3R/2}}$$

$$\leq N\sum_{|\alpha|\leq k_0+1}|D^\alpha f|_{\delta/2,\delta;Q_{2R}}+N\sum_{|\alpha|\leq k_0}|D^\alpha u|_{1+\delta/2,2+\delta;Q_{3R/2}}$$

$$\leq N\sum_{|\alpha|\leq k_0+1}|D^\alpha f|_{\delta/2,\delta;Q_{2R}}+N|u|_{0;Q_{2R}}.$$

This gives (8.12.1) for $k=k_0+1$, and the theorem is proved.

COROLLARY 8.12.2. *Let the assumptions of Theorem 8.12.1 be satisfied for* $R=\infty$, *and let* $c\leq-\lambda$ *for a constant* $\lambda>0$. *Then for* $Q=(-\infty,0)\times\mathbb{R}^d$ *we have*

$$\sum_{|\alpha|\leq k}|D^\alpha u|_{1+\delta/2,2+\delta;Q}\leq N\sum_{|\alpha|\leq k}|D^\alpha(Lu-u_t)|_{\delta/2,\delta;Q},$$

where the constant N *depends only on* $K,\delta,k,d,\kappa,\lambda$.

Indeed, in this situation it suffices to apply (8.12.1) to $z+Q_1$, $z+Q_2$ with z running through Q and to notice that $|u|_{0;Q}\leq\lambda^{-1}|Lu-u_t|_{0;Q}$.

REMARK 8.12.3. Theorem 8.12.1 and Corollary 8.12.2 related to parabolic equations *contain* Theorem 4.3.1 and Theorem 7.1.1 related to second-order elliptic equations.

In the case of Theorem 7.1.1 this is obvious. To get Theorem 4.3.1 we apply Corollary 8.12.2 to functions independent of t. Then we obtain (4.3.1) with $z = 1$. Standard dilations prove (4.3.1) for larger values of z, and for smaller $z > 0$ we just repeat the short argument from the proof of Theorem 4.3.1.

EXERCISE 8.12.4. Assume that for an $R > 0$ and a given integer $k \geq 0$ we have $|D_t^r a|_{\delta/2,\delta;Q_{2R}}$, $|D_t^r b|_{\delta/2,\delta;Q_{2R}}$, $|D_t^r c|_{\delta/2,\delta;Q_{2R}} \leq K$, for any $r \leq k$. Assume that $u \in C^{1+\delta/2,2+\delta}(Q_{2R})$ and $D_t^r(Lu - u_t) \in C^{\delta/2,\delta}(Q_{2R})$ for $r \leq k$. Under these conditions prove that $D_t^r u \in C^{1+\delta/2,2+\delta}(Q_R)$ if $r \leq k$ and that there is a constant N depending only on $R, K, \delta, k, d, \kappa$ such that

$$\sum_{r\leq k} |D_t^r u|_{1+\delta/2,2+\delta;Q_R} \leq N \sum_{r\leq k} |D_t^r(Lu - u_t)|_{\delta/2,\delta;Q_{2R}} + N|u|_{0;Q_{2R}}.$$

EXERCISE 8.12.5 (better regularity of solutions of nonlinear equations). This exercise is similar to Exercise 4.2.4. In notation from Exercise 4.2.4 for $m = 2$ let $F(x, \xi^{(\alpha)})$ be a real-valued $C^{1+\delta}(\mathbb{R}^d \times \mathbb{R}^{d_1})$-function, and let $u \in C^{1+\delta/2,2+\delta}(\mathbb{R}^{d+1})$ be a solution of the equation $F(x, D^\alpha u(x)) = u_t$ on $\mathbb{R}^{d+1}$. Assume that the equation is *parabolic on* u, which means that if we define $a^\beta(x) = F_{\xi^{(\beta)}}(x, D^\alpha u(x))$ and $L = \sum_{|\alpha|\leq 2} a^\alpha D^\alpha$, then L is a second-order elliptic operator. Prove that in this situation $u_x \in C^{1+\delta/2,2+\delta}(\mathbb{R}^{d+1})$.

8.13. Hints to exercises

8.1.10. Take $\varepsilon > 0$ and for $w = u(t+\varepsilon, x) - u(t,x)$ find $Lw - w_t$.
8.1.12. When x is close to Γ_i prove the inequality for any $\lambda \geq 0$. If x is not close to Γ_i, observe that $\psi_i(x)$ is not close to zero and choose λ large enough.
8.1.13. Take ϕ_i from Exercise 8.1.12 and define

$$M = \limsup_{Q\ni z\to\partial_{tx}Q} u_+(z).$$

For $\gamma > 0$ consider the function $u_\gamma(t,x) := u(t,x) - Mh(\gamma)\sum_i \phi_i(t+\gamma, x) - \gamma(T-t)^{-1}$, where $h(\gamma) = \gamma^{\varepsilon\kappa}$. Show that if $\sup u_\gamma > 0$, then u_γ takes its maximum value over $\bar{Q} \setminus \partial_{tx}Q$ inside Q.
8.1.15. In the case $T < \infty$ follow the hint to Exercise 2.6.9, but also subtract $\varepsilon e^{-t}\cos x^d$.
8.1.16. Take the same function v as in the hint to Exercise 2.6.10.
8.1.17. (i) Consider $u(t,x) - u(t-\varepsilon, x)$, where $\varepsilon > 0$. (ii) Consider $u(t, x+y) - u(t,x)$, where y is fixed and $y^d = 0$. (iii) Owing to (i) and (ii), one needs only to show that $|u(t, x+\varepsilon e_d) - u(t,x)| \leq N\varepsilon^\delta$. From the maximum principle it is clear that it suffices to prove this only for $x^d = 0$. Without loss of generality one may consider $u(0,x) - u(0,0)$. Now as in the hint to Exercise 2.6.11 for $t \leq 0, x \in \mathbb{R}^d_+$ consider $u(t,x) - u(0,0) - v(x) - \phi(t,x)$, where $\phi(t,x) = ((x^d)^2 - t)^{\delta/2} + c(x^d)^\delta$ and the constant $c > 0$ is so large that $\Delta\phi - \phi_t \leq 0$ in $(-\infty, 0) \times \mathbb{R}^d_+$.

Alternatively, one can get all assertions from the integral representation which will be derived and used in Sec. 9.4.

8.1.18. To prove the first inequality one considers the function $u(t,x+y)-u(t,x)$, where y is fixed. A similar argument shows that the second inequality needs to be proved only for $s=0$. Fix a $\varepsilon>0$ and define $\mu(\varepsilon)$ so that $\varepsilon|x|^2+\mu(\varepsilon)\geq|x|^\delta$ for any $x\in\mathbb{R}^d$. Consider the function $u(t,x)-u(0,0)-\varepsilon(|x|^2+t2d)$ and by the maximum principle prove that it is less than $\mu(\varepsilon)$. Then take $x=0$ and minimize with respect to ε.

Another way to do this exercise is to apply an explicit formula for u from Sec. 9.1.

8.1.19. Observe that with some bounded b we have $f=\Delta u+b^iu_{x^i}-u-u_t$, and also see the hint to Exercise 4.3.6.

8.1.20. Apply the operator $L-\partial/\partial t$ to the function $u-v/t$, bearing in mind that $u\leq 1$ so that there is no need to consider the points where $v(x)\geq t$.

8.1.21. Define $T=2\sup_\Omega v$ and observe that $m(t):=\sup_x u_+(t,x)$ decreases and $m(T)\leq 1/2$. Conclude that $m(2T)\leq(1/2)^2$,..., $m(nT)\leq(1/2)^n$.

8.1.22. Notice that for $v_0=(1+|x|^2)\exp(2Kt)$ one has $Lv_0-v_{0t}\leq 0$.

8.1.23. Check that for any $\varepsilon>0$ the function $v_0=(\varepsilon-t)^{-d/2}\exp(|x|^2/(4\varepsilon-4t))$ satisfies $\Delta v_0-v_{0t}=0$ for $t\in(0,\varepsilon)$, $x\in\mathbb{R}^d$. Then notice that the function $\bar{u}:=u/v_0$ tends to zero as $|x|\to\infty$, $t\in(0,\varepsilon)$ if ε is small enough, and also $L'\bar{u}-\bar{u}_t=0$, where the operator L' is defined by the formula $L'w=v_0^{-1}[\Delta(v_0w)-w\Delta v_0]$. By using Corollary 8.1.3 conclude that $\bar{u}=0$, $u=0$ for $t\in[0,\varepsilon]$, and then repeat the argument for $u(t+\varepsilon,x)$ instead of u.

8.2.3. Reduce $\partial(E_\eta\setminus B_{z_0})$ to ∂B_{z_0} and then prove that the right-hand side solves the equation $u_t=Au$ and $u(0)=f$. In the proof of the latter make the circle over which we are integrating expand to infinity.

8.2.5. Multiply the right-hand side by $z-A$ and integrate by parts.

8.2.6. Observe that the integral equals

$$(A-z)\int_0^\infty\frac{1}{t^{3/2}}\int_0^t e^{(A-z)s}\,ds\,dt=2(A-z)\int_0^\infty\frac{1}{t^{1/2}}e^{(A-z)t}\,dt,$$

so that its square equals

$$4(A-z)^2\int_0^\infty\int_0^\infty\frac{1}{t^{1/2}}\frac{1}{s^{1/2}}e^{(A-z)(t+s)}\,dsdt.$$

Now change the variables and use Exercise 8.2.5.

8.3.4. First prove that the integral converges. Then notice that it suffices to take f in the form $zv-Lv$, where $v\in C^{2+\delta}(\mathbb{R}^d)$. After this prove that $T_tLv=LT_tv=(\partial T_tv)/(\partial t)$ and integrate by parts.

8.3.6. For s fixed write down the equation for $T_tT_sf(x)$ as a function of (t,x) and use uniqueness.

8.4.8. Observe that $G_\lambda\leq G_0$, then prove that $G_0*I_{Q_1}$ is bounded and finally use dilations. One can take $N(d)=1$.

8.5.8. Consider $(x-\mu)_+^\delta$ where μ is a positive parameter.

8.5.10. Proceed as in the hint to Exercise 3.3.4 after observing that by definition $|u^{(\varepsilon)}(z_1)-u^{(\varepsilon)}(z_2)|\leq\rho^\gamma(z_1,z_2)\varepsilon^{\delta-\gamma}[u]''_{\delta/2,\delta}$.

8.6.2. Use parabolic dilations, that is, transformation $(t,x)\to(c^2t,cx)$.

8.7.4. Define $v(t,x)=u(t\lambda^{-1},x\lambda^{-1/2})$, $g=\Delta v-v_t-v$ and prove that $g(t,x)=\lambda^{-1}f(t\lambda^{-1},x\lambda^{-1/2})$ and

$$|v|_0 = |u|_0, \quad |v_x|_0 = \lambda^{-1/2}|u_x|_0, \quad |v_{xx}|_0 + |v_t|_0 = \lambda^{-1}(|u_{xx}|_0 + |u_t|_0),$$

$$[v]_{1+\delta/2,2+\delta} = \lambda^{-1-\delta/2}[u]_{1+\delta/2,2+\delta}, \quad |g|_0 = \lambda^{-1}|f|_0, \quad [g]_{\delta/2,\delta} = \lambda^{-1-\delta/2}[f]_{\delta/2,\delta}.$$

One can also get the result by considering $u(t,x)\exp(i\sqrt{\lambda}x^{d+1})$ instead of $u(t,x)\exp(ix^{d+1})$ in the proof of Theorem 8.7.2.

8.8.2. Use the fact that ε in Theorem 8.8.1 is an arbitrary positive number.

8.8.4. Apply *the Hadamard formula*:

$$F(x,\xi) - F(x,\eta) = (\xi^i - \eta^i)\int_0^1 F_{\xi^i}(x, t\xi + (1-t)\eta)\,dt.$$

8.8.6. Follow the proof of Theorem 8.8.1, but do not use $|a+b+c| \le |a|+|b|+|c|$ too soon. For instance

$$|[u(t,x+\varepsilon e_i) - u(s,x+\varepsilon e_i)] - [u(t,x) - u(s,x)]| = |t-s|\cdot|u_t(\theta,x+\varepsilon e_i) - u_t(\theta,x)|.$$

8.8.8. While estimating $|u(t,x)-u(s,x)|$ or $|u(t,x)-u(t,y)|$ consider u as a function of only t or x.

8.8.9. Observe that $[u^{(\varepsilon)}]_{1+\delta/2,2+\delta} \le N[f^{(\varepsilon)}]_{\delta/2,\delta}$ where $f = \Delta u - u_t$. Also notice that by (elliptic) interpolation inequalities

$$[f]_{\delta/2,\delta} \le N(|f_t|_0^{\delta/2}|f|_0^{1-\delta/2} + |f_x|_0^{\delta}|f|_0^{1-\delta}).$$

Next, Hölder's inequality easily implies that

$$|f^{(\varepsilon)}|_0 \le N\varepsilon^{-(d+2)/p}||f||_{L_p(\mathbb{R}^{d+1})}, \quad |f_t^{(\varepsilon)}|_0 \le N\varepsilon^{-(d+2)/p-2}||f||_{L_p(\mathbb{R}^{d+1})},$$

$$|f_x^{(\varepsilon)}|_0 \le N\varepsilon^{-(d+2)/p-1}||f||_{L_p(\mathbb{R}^{d+1})}.$$

Combining these results yields

$$[f^{(\varepsilon)}]_{\delta/2,\delta} \le N\varepsilon^{-\delta-(d+2)/p}||f||_{L_p(\mathbb{R}^{d+1})}.$$

Now in Exercise 8.5.10 take $\gamma = 1$, apply Exercise 8.8.8 and notice that $[u^{(\varepsilon)}]_{\delta/2,\delta} \le [u]_{\delta/2,\delta}$.

8.9.3. As in Exercise 8.7.4 use parabolic dilations, observing that under dilations the coefficients of L are transformed into other functions with *smaller* norms $|\cdot|_{\delta/2,\delta}$.

8.9.5. Observe that by Theorem 8.7.2 we have $|u|_{1+\delta/2,2+\delta} \le N|f - F(\operatorname{grad} u)|_{\delta/2,\delta}$. Then use Exercises 8.8.4 and 8.1.19.

8.10.4. Notice that from Exercise 8.9.5 one has an a priori estimate, and then proceed as in the hint to Exercise 4.3.9.

8.10.5. First let $k > 0$ and observe that for any given f, owing to Exercise 8.9.5, one can replace $|\operatorname{grad} u|^{k+1}$ by $F(\operatorname{grad} u)$ with F satisfying the conditions from Exercise 8.10.4 and, moreover, in this way obtaining an *equivalent* equation. For $k = 0$ use a passage to the limit.

8.12.5. Proceed as in the hint to Exercise 4.2.4 by using Exercise 8.8.5 and first prove that $u_x \in C^{1+\delta_1/2,2+\delta_1}(\mathbb{R}^{d+1})$ with small $\delta_1 > 0$. Then from Exercise 8.8.6

obtain that first derivatives u_{xx} of u_x are Lipschitz continuous with respect to x and $|u_{xx}(t,x) - u_{xx}(s,x)| \leq N|t-s|^{1/2}$. After this again use Exercise 8.8.5.

CHAPTER 9

Boundary-Value Problems for Parabolic Equations in Half Spaces

In this chapter we continue to study elliptic second–order operators L with real coefficients as in (8.0.2). We always assume that L is nondegenerate with constant of ellipticity $\kappa > 0$. Also let $|a, b, c|_{\delta/2,\delta} \le K$, where $\delta \in (0,1)$ and K are fixed constants. We study the Dirichlet problem for parabolic equations in half spaces.

Here we deal with only two kinds of half spaces in $\mathbb{R}^{d+1}$: $\mathbb{R}_+ \times \mathbb{R}^d = \{z = (t,x) : t > 0\}$ when we have to solve the Cauchy problem and $\mathbb{R}^d \times \mathbb{R}_+ = \{z = (t, x', x^d) : x^d > 0\}$, which corresponds to the boundary-value problem. In a sense these are the only half spaces of interest. Indeed, it turns out that the Dirichlet problem for a parabolic equation $Lu - u_t = f$ in a spacelike half space $\{(t,x) : t > h \cdot x\}$, where $h \in \mathbb{R}^d \setminus \{0\}$, can be reduced to the same problem in $\mathbb{R}^d \times \mathbb{R}_+$. To see this without loss of generality, assume that $h^d \ne 0$. Then it suffices to change the independent variables, keeping $t, x^1, ..., x^{d-1}$ unchanged and taking $t - h \cdot x$ as the new variable x^d.

Actually, to make the results look more general, we consider $(0,T) \times \mathbb{R}^d$ and $(-\infty, T) \times \mathbb{R}^d_+$, where $T \in (0, \infty]$, instead of $\mathbb{R}_+ \times \mathbb{R}^d$ and $\mathbb{R}^d \times \mathbb{R}_+$.

9.1. The Cauchy problem for the heat equation

Let $T \in (0, \infty]$, $Q = (0,T) \times \mathbb{R}^d$.

EXERCISE 9.1.1. Let $u \in C^{1+\delta/2,2+\delta}(Q)$ and $\Delta u = u_t$ in Q. On the basis of Exercise 8.1.18 prove that $|u|_{1+\delta/2,2+\delta;Q} \le N(d)|u(0,\cdot)|_{2+\delta}$.

In the following theorem we prove a similar estimate and the solvability of the Cauchy problem for the heat equation.

THEOREM 9.1.2. *Let $g \in C^{2+\delta}(\mathbb{R}^d)$ and $f \in C^{\delta/2,\delta}(Q)$. Then there exists a unique function $u \in C^{1+\delta/2,2+\delta}(Q)$ such that*

$$\Delta u - u_t - u = f \tag{9.1.1}$$

in Q and $u(0,x) = g(x)$ in $\mathbb{R}^d$. Moreover, there exists a constant N depending only on d, δ such that

$$|u|_{1+\delta/2,2+\delta;Q} \le N\Big(|f|_{\delta/2,\delta;Q} + |g|_{2+\delta}\Big). \tag{9.1.2}$$

Proof. First, notice that uniqueness follows from the maximum principle. Further, we can continue f for negative t by $f(t,x) = f(0,x)$, and by Theorem 8.10.1 for $Q' = (-\infty, T) \times \mathbb{R}^d$ we can find a unique function of class $C^{1+\delta/2,2+\delta}(Q')$ satisfying equation (9.1.1) in Q'. From Theorem 8.10.1 we also have an estimate of $C^{1+\delta/2,2+\delta}(Q')$–norm of this function through $C^{\delta/2,\delta}(Q')$–norm of the extended f.

The last norm coincides with $|f|_{\delta/2,\delta;Q}$, and this argument implies that we need to prove the present theorem only for $f \equiv 0$.

Define, $u(t,x) = G_1(t,x) * g(x)$, where G_λ is introduced in (8.4.1) and the convolution is performed with respect to the x variable. Since every derivative of $G_1(t,x)$ is bounded by a summable function whenever t is bounded away from zero, we see that $u(t,x)$ is infinitely differentiable in Q and every derivative of $u(t,x)$ is bounded whenever t is bounded away from zero. In particular, $u \in C^{1+\delta/2,2+\delta}(Q(\varepsilon))$ for any $\varepsilon > 0$, where $Q(\varepsilon) = (\varepsilon, 0) + Q$. Moreover, from (8.4.2) it follows that $\Delta u - u_t - u = 0$ in Q. Also

$$u(t,x) = e^{-t}\frac{1}{(4\pi)^{d/2}}\int_{\mathbb{R}^d} g(x + z\sqrt{t})e^{-|z|^2/4}\,dz, \tag{9.1.3}$$

$$\lim_{t\downarrow 0} u(t,x) = g(x)\frac{1}{(4\pi)^{d/2}}\int_{\mathbb{R}^d} e^{-|z|^2/4}\,dz = g(x).$$

Furthermore, from (9.1.3) we see that $|u|_{0;Q} \le |g|_0$,

$$|u_{xx}(t,x)| \le G_1(t,x) * |g_{xx}(x)| \le N|g|_{2+\delta},$$

$$\frac{|u_{xx}(t,x) - u_{xx}(t,y)|}{|x-y|^\delta} \le e^{-t}\frac{1}{(4\pi)^{d/2}}\int_{\mathbb{R}^d}\frac{|g_{xx}(x+z\sqrt{t}) - g_{xx}(y+z\sqrt{t})|}{|x-y|^\delta}e^{-|z|^2/4}\,dz$$

$$\le N|g|_{2+\delta}.$$

To estimate Hölder's constant of u_{xx} with respect to t, we observe that if $t \ge s+1$, then $e^{-t}|\sqrt{t} - \sqrt{s}|^\delta \le e^{-t}t^{\delta/2} \le N \le N|t-s|^{\delta/2}$. If $s+1 \ge t \ge s \ge 0$, then from $|\sqrt{t} - \sqrt{s}| \le \sqrt{t-s}$ we get $|\sqrt{t} - \sqrt{s}|^\delta \le |t-s|^{\delta/2}$. Therefore, $e^{-t}|\sqrt{t} - \sqrt{s}|^\delta \le N|t-s|^{\delta/2}$ for any $t \ge s \ge 0$ and

$$\frac{|u_{xx}(t,x) - u_{xx}(s,x)|}{|t-s|^{\delta/2}} \le \frac{|e^{-t} - e^{-s}|}{|t-s|^{\delta/2}}|g_{xx}|_{0;Q}$$

$$+ e^{-t}\frac{1}{(4\pi)^{d/2}}\int_{\mathbb{R}^d}\frac{|g_{xx}(x+z\sqrt{t}) - g_{xx}(x+z\sqrt{s})|}{|t-s|^{\delta/2}}e^{-|z|^2/4}\,dz$$

$$\le N|g|_{2+\delta} + N|g|_{2+\delta}\frac{1}{(4\pi)^{d/2}}\int_{\mathbb{R}^d}|z|^\delta e^{-|z|^2/4}\,dz \le N|g|_{2+\delta}.$$

Thus, $[u_{xx}]_{\delta/2,\delta;Q} \le N|g|_{2+\delta}$. From the equation $u_t = \Delta u - u$ we see that $[u_t]_{\delta/2,\delta;Q(\varepsilon)} \le N(|g|_{2+\delta} + [u]_{\delta/2,\delta;Q(\varepsilon)})$. Now remember that $u \in C^{1+\delta/2,2+\delta}(Q(\varepsilon))$ and apply interpolation inequalities. Then we conclude

$$|u|_{1+\delta/2,2+\delta;Q(\varepsilon)} \le N(|g|_{2+\delta} + [u]_{\delta/2,\delta;Q(\varepsilon)}) + |u|_{0;Q} \le N|g|_{2+\delta} + N[u]_{\delta/2,\delta;Q(\varepsilon)}$$

$$\le N|g|_{2+\delta} + (1/2)|u|_{1+\delta/2,2+\delta;Q(\varepsilon)} + N|u|_{0;Q} \le N|g|_{2+\delta} + (1/2)|u|_{1+\delta/2,2+\delta;Q(\varepsilon)}.$$

By collecting like terms we get (9.1.2) with $Q(\varepsilon)$ instead of Q on the left. It remains only to make an obvious passage to the limit. The theorem is proved.

EXERCISE 9.1.3. One might prefer to deal with the equation $\Delta u - u_t = f$ instead of (9.1.1). In connection with this show that if $T < \infty$ and $g \in C^{2+\delta}(\mathbb{R}^d)$ and $f \in C^{\delta/2,\delta}(Q)$, then there exists a unique function $u \in C^{1+\delta/2,2+\delta}(Q)$ such that $\Delta u - u_t = f$ in Q and $u(0,x) = g(x)$ in $\mathbb{R}^d$. Also show that there exists a constant $N = N(d,\delta)$ such that

$$[u]_{1+\delta/2,2+\delta;Q} + T^{-\delta/2}(|u_{xx}|_{0;Q} + |u_t|_{0;Q}) + T^{-(1+\delta)/2}|u_x|_{0;Q}$$

$$+T^{-(2+\delta)/2}|u|_{0;Q} \leq N\Big([f]_{\delta/2,\delta;Q} + T^{-\delta/2}|f|_{0;Q}$$

$$+[g]_{2+\delta} + T^{-\delta/2}|g_{xx}|_{0;Q} + T^{-(1+\delta)/2}|g_x|_{0;Q} + T^{-(2+\delta)/2}|g|_{0;Q}\Big).$$

EXERCISE 9.1.4. By letting $T \to \infty$ in Exercise 9.1.3 one obtains that for $Q = (0,\infty) \times \mathbb{R}^d$ and any $u \in C^{1+\delta/2,2+\delta}(Q)$

$$[u]_{1+\delta/2,2+\delta;Q} \leq N\Big([\Delta u - u_t]_{\delta/2,\delta;Q} + [u(0,\cdot)]_{2+\delta}\Big). \tag{9.1.4}$$

Prove that (9.1.4) holds with $N = N(\delta,d)$ if we replace Q with $(0,T) \times \mathbb{R}^d$ for any $T > 0$.

EXERCISE 9.1.5. Deduce from (9.1.4) only that with a constant $N = N(d,\delta)$ for any $\lambda > 0$ and $u \in C^{1+\delta/2,2+\delta}(Q)$ we have

$$[u]_{1+\delta/2,2+\delta;Q} + \lambda^{\delta/2}(|u_{xx}|_{0;Q} + |u_t|_{0;Q}) + \lambda^{(1+\delta)/2}|u_x|_{0;Q} + \lambda^{1+\delta/2}|u|_{0;Q}$$

$$\leq N([f]_{\delta/2,\delta;Q} + \lambda^{\delta/2}|f|_{0;Q} + [u(0,\cdot)]_{2+\delta} + \lambda^{1+\delta/2}|u(0,\cdot)|_0), \tag{9.1.5}$$

where $f := \Delta u - u_t - \lambda u$.

In the following two exercises we propose to combine the result of this section, giving an explicit solution of the Cauchy problem for the heat equation, and the ideas from Secs. 8.2 and 8.3 bearing on the semigroup approach.

EXERCISE 9.1.6. Using Corollary 8.3.1 and formula (8.3.3), derive again (1.6.2).

EXERCISE 9.1.7. In Sec. 5.5 we have introduced the operator $\mathcal{K}_1 : C^{1+\delta}(\mathbb{R}^{d-1}) \to C^{\delta}(\mathbb{R}^{d-1})$. To reflect that the operator depends on d, write $\mathcal{K}_{1,d}$. Prove that if $g \in C_0^\infty(\mathbb{R}^d)$, then

$$-\mathcal{K}_{1,d+1}g(x) = c\int_0^\infty \frac{1}{t^{3/2}}[G_1(t,x) * g(x) - g(x)]\,dt,$$

where c is the constant from Exercise 8.2.6. Put another way, prove that the result of Exercise 8.2.6 holds for $A = \Delta$, $z = 1$.

9.2. The Cauchy problem for general parabolic equations

Here we consider the Cauchy problem for equation (8.0.1) with a general operator L. Let $T \in (0,\infty]$, $Q = (0,T) \times \mathbb{R}^d$.

LEMMA 9.2.1. *Assume that the coefficients a^{ij} are constant and $b^i = c = 0$. Then there exists a constant N depending only on d, κ, δ, K such that for any $u \in C^{1+\delta/2,2+\delta}(Q)$ we have*

$$|u|_{1+\delta/2,2+\delta;Q} \le N\Big(|Lu - u_t - u|_{\delta/2,\delta;Q} + |u(0,\cdot)|_{2+\delta}\Big). \tag{9.2.1}$$

To prove this lemma it suffices to apply Theorem 9.1.2 after having made an appropriate change of coordinates as in the proof of Lemma 8.9.1.

Next we prove (9.2.1) for general operators L.

THEOREM 9.2.2. *There exists a constant N depending only on d, κ, δ, K such that estimate (9.2.1) holds for any $u \in C^{1+\delta/2,2+\delta}(Q)$.*

Owing to Lemma 9.2.1, to prove this theorem all we need to do is just to repeat the proof of Theorem 8.9.2 with *only one* change: on the right in the inequalities of that proof we will now have the additional term

$$N|(u\xi)(0,\cdot)|_{2+\delta} \le N(\gamma)|u(0,\cdot)|_{2+\delta}.$$

On the basis of Theorems 9.1.2 and 9.2.2 as in the proof of Theorem 8.10.1 we arrive at the following result.

THEOREM 9.2.3. *In addition to our general hypotheses on L assume that $c \le -\lambda$ where the constant $\lambda > 0$. Let $g \in C^{2+\delta}(\mathbb{R}^d)$ and $f \in C^{\delta/2,\delta}(Q)$. Then there exists a unique function $u \in C^{1+\delta/2,2+\delta}(Q)$ such that it satisfies equation (8.0.1) in Q and $u(0,x) = g(x)$ in $\mathbb{R}^d$. Moreover, there exists a constant N depending only on $d, \delta, K, \kappa, \lambda$ such that*

$$|u|_{1+\delta/2,2+\delta;Q} \le N\Big(|f|_{\delta/2,\delta;Q} + |g|_{2+\delta}\Big). \tag{9.2.2}$$

EXERCISE* 9.2.4 (cf. Exercise 8.9.3). Prove that there exists a constant N depending only on d, κ, δ, K such that for any $\lambda \ge 1$ and $u \in C^{1+\delta/2,2+\delta}(Q)$ we have

$$|u|_{1+\delta/2,2+\delta;Q} \le N\Big([f]_{\delta/2,\delta;Q} + \lambda^{\delta/2}|f|_{0;Q} + [u(0,\cdot)]_{2+\delta} + \lambda^{1+\delta/2}|u(0,\cdot)|_0\Big),$$

$$|u|_0 \le \lambda^{-1}|f|_{0,Q} + |u(0,\cdot)|_0,$$

where $f = Lu - u_t - \lambda u$.

EXERCISE 9.2.5. By imitating the proof of Theorem 8.11.1 and using (9.2.2) prove that if $z \in \mathbb{R}^{d+1}$ and $u \in C^{1+\delta/2,2+\delta}(Q_{3R}(z)\cap Q)$ and $u = 0$ on $Q_{3R}(z)\cap\partial' Q$, then

$$|u|_{1+\delta/2,2+\delta;Q_R(z)\cap Q} \le N\Big(|Lu - u_t|_{\delta/2,\delta;Q_{2R}(z)\cap Q} + |u|_{0;Q_{2R}(z)\cap Q}\Big),$$

where N depends only on R, d, δ, K, κ.

9.3. An equivalent norm in parabolic Hölder spaces

Take $T \in (-\infty, \infty]$ and $\delta \in (0,1)$. Also take the operators $\partial_{h,j}$, ∂_h^α from Sec. 5.1 and define $\partial_{h,t}^0 u \equiv u$,

$$\partial_{h,t}^1 u(t,x) = u(t,x) - u(t-h^2,x), \quad \partial_{h,t}^{3/2} u(t,x) = (\partial_{h,t}^1)^2 u(t,x).$$

EXERCISE 9.3.1. By using Exercise 8.1.17 prove that if $Q = (-\infty, T) \times \mathbb{R}^d_+$ and $u \in C^{1+\delta/2,2+\delta}(Q)$ and $\Delta u = u_t$ in Q, then $[u_t]_{\delta/2,\delta;Q} \leq N[g_t]_{\delta/2,\delta,\partial' Q}$ and $[u_{x^i x^j}]_{\delta/2,\delta;Q} \leq N[g_{x^i x^j}]_{\delta/2,\delta,\partial' Q}$ for $i,j = 1,...,d-1$ and $[u_{x^d x^d}]_{\delta/2,\delta;Q} \leq N[g]_{1+\delta/2,2+\delta,\partial' Q}$ where $N = N(d,\delta)$ and g is the restriction of u on $\partial' Q$.

This exercise shows that we intend to use Corollary 9.3.3 in the same way as a similar fact was used in elliptic theory.

THEOREM 9.3.2. *Let Q be $(-\infty, T) \times \mathbb{R}^d_+ = \{(t,x) : t < T, x^d > 0\}$ or $(-\infty, T) \times \mathbb{R}^d$. For a function u defined in Q denote*

$$[u]'_{1+\delta/2,2+\delta;Q} = \sup_{\beta=0,1,3/2} \sup_{2\beta+|\alpha|=3} \sup_{h>0} \sup_{z\in Q} \frac{1}{h^{2+\delta}} |\partial_{h,t}^\beta \partial_h^\alpha u(z)|.$$

Then there exists a constant $N = N(d,\delta)$ such that for any $u \in C^{1+\delta/2,2+\delta}(Q)$ we have

$$[u]_{1+\delta/2,2+\delta;Q} \leq N[u]'_{1+\delta/2,2+\delta;Q}, \quad [u]'_{1+\delta/2,2+\delta;Q} \leq N[u]_{1+\delta/2,2+\delta;Q}. \tag{9.3.1}$$

Proof. Fix t and consider $u(t,x)$ as a function of x. Then from Lemma 5.1.2 we get that for $|\alpha| = 3$

$$|\partial_h^\alpha u(t,x)| \leq Nh^{2+\delta}[u_{xx}(t,\cdot)]_{\delta;\Omega} \leq Nh^{2+\delta}[u]_{1+\delta/2,2+\delta;Q}, \tag{9.3.2}$$

where Ω is $\mathbb{R}^d_+$ or $\mathbb{R}^d$. If instead we fix x, then for $\beta = 3/2$ from Lemma 5.1.2 we get that

$$|\partial_{h,t}^\beta u(t,x)| \leq Nh^{2+\delta}[u_t(\cdot,x)]_{\delta/2;(-\infty,T]} \leq Nh^{2+\delta}[u]_{1+\delta/2,2+\delta;Q}. \tag{9.3.3}$$

Furthermore, for $|\alpha| = 1$ there is an integer j such that $\alpha = e_j$, and by the mean value theorem we have

$$|\partial_{h,t}^1 \partial_h^\alpha u(t,x)| = h^2 |\partial_{h,j} u_t(\theta, x)|,$$

where $t - h^2 < \theta < t$. Therefore,

$$|\partial_{h,t}^1 \partial_h^\alpha u| \leq h^{2+\delta}[u_t]_{\delta/2,\delta;Q}, \tag{9.3.4}$$

and this along with (9.3.2) and (9.3.3) proves the second assertion in (9.3.1).

The proof of the first one we start by noticing that the inequality

$$|u_{xx}(t,x) - u_{xx}(t,y)| \leq N|x-y|^\delta [u]'_{1+\delta/2,2+\delta;Q} \tag{9.3.5}$$

follows from the first inequality in (5.1.1). In estimating Hölder's constant of u_{xx} with respect to t we proceed as in the proof of Lemma 5.1.2. We have

$$|u_{x^i x^j}(t,x) - u_{x^i x^j}(t-h^2,x)| \leq |u_{x^i x^j}(t,x) - n^2 h^{-2} \partial_{h/n,i} \partial_{h/n,j} u(t,x)|$$

$$+n^2h^{-2}\sum_{r=0}^{n^2-1}|\partial^1_{h/n,t}\partial_{h/n,i}\partial_{h/n,j}u(t-rh^2n^{-2},x)|$$

$$+|u_{x^ix^j}(t-h^2,x)-n^2h^{-2}\partial_{h/n,i}\partial_{h/n,j}u(t-h^2,x)|.$$

From the mean value theorem it follows that the first and the last terms on the right are less than $N(h/n)^\delta[u_{xx}]_{\delta/2,\delta;Q}$. Also, the middle term is less than

$$2n^2h^{-2}\sum_{r=0}^{n^2-1}\sup_{z\in Q}|\partial^1_{h/n,t}\partial_{h/n,i}u(z)|\le 2n^{2-\delta}h^\delta[u]'_{1+\delta/2,2+\delta;Q}.$$

Therefore,

$$\sup_{t,x,h}\frac{|u_{x^ix^j}(t,x)-u_{x^ix^j}(t-h^2,x)|}{h^\delta}\le Nn^{-\delta}[u_{xx}]_{\delta/2,\delta;Q}+2n^{2-\delta}[u]'_{1+\delta/2,2+\delta;Q},$$

which together with (9.3.5) shows that $[u_{xx}]_{\delta/2,\delta;Q}\le Nn^{-\delta}[u_{xx}]_{\delta/2,\delta;Q}+N(1+n^{2-\delta})[u]'_{1+\delta/2,2+\delta;Q}$. Since n is arbitrary and N is independent of n, we have estimated $[u_{xx}]_{\delta/2,\delta;Q}$ through $[u]'_{1+\delta/2,2+\delta;Q}$. It remains to estimate the Hölder constant of u_t.

By Lemma 5.1.2 for any x and for $t<T, h>0$

$$|u_t(t,x)-u_t(t-h^2,x)|\le Nh^\delta\sup_{s>0,r<T}s^{-1-\delta/2}|u(r,x)-2u(r-s,x)+u(r-2s,x)|$$

$$=Nh^\delta\sup_{s>0,r<T}s^{-2-\delta}|\partial^{3/2}_{s,t}u(r,x)|\le Nh^\delta[u]'_{1+\delta/2,2+\delta;Q}.$$

Furthermore, for any $i=1,...,d$ and integer n

$$|u_t(t,x+he_i)-u_t(t,x)|\le|u_t(t,x+he_i)-n^2h^{-2}\partial^1_{h/n,t}u(t,x+he_i)|$$

$$+n^2h^{-2}\sum_{r=0}^{n-1}|\partial_{h/n,i}\partial^1_{h/n,t}u(t,x+e_irh/n)|+|u_t(t,x)-n^2h^{-2}\partial^1_{h/n,t}u(t,x)|.$$

As above, this allows us to get

$$|u_t(t,x+he_i)-u_t(t,x)|\le 2h^\delta n^{-\delta}[u_t]_{\delta/2,\delta;Q}+n^{1-\delta}h^\delta[u]'_{1+\delta/2,2+\delta;Q}.$$

We finally see that $[u_t]_{\delta/2,\delta;Q}\le Nn^{-\delta}[u_t]_{\delta/2,\delta;Q}+N(1+n^{1-\delta})[u]'_{1+\delta/2,2+\delta;Q}$ for any n with N independent of n, which yields $[u_t]_{\delta/2,\delta;Q}\le N[u]'_{1+\delta/2,2+\delta;Q}$. The theorem is proved.

Inequalities (9.3.2), (9.3.3) and (9.3.4) and Theorems 9.3.2 and 5.1.3 give us the following useful assertion.

COROLLARY 9.3.3. *Let Q be $(-\infty,T)\times\mathbb{R}^d_+$ or $(-\infty,T)\times\mathbb{R}^d$ and take $u\in C^{1+\delta/2,2+2+\delta}(Q)$. Then*

$$[u]_{1+\delta/2,2+\delta;Q}$$

$$\le N(d,\delta)\max\{[u_t]_{\delta/2,\delta;Q},[u_{x^ix^j}]_{\delta/2,\delta;Q},[u_{x^dx^d}]_{\delta/2,\delta;Q}:i,j=1,...,d-1\}.$$

9.4. Boundary-value problem for general operators in half spaces

Let $T \in (-\infty, \infty]$ and $Q = (-\infty, T) \times \mathbb{R}^d_+ = (-\infty, T) \times \{(x', x^d) : x' \in \mathbb{R}^{d-1}, x^d > 0\}$.

THEOREM 9.4.1. *In addition to our general hypotheses on L assume that $c \le -\lambda$ where the constant $\lambda > 0$. Let $g = g(t, x')$ be a function of class $C^{1+\delta/2, 2+\delta}(\partial' Q)$ and $f \in C^{\delta/2,\delta}(Q)$. Then there exists a unique function $u \in C^{1+\delta/2,2+\delta}(Q)$ satisfying equation (8.0.1) in Q and such that $u(t, x', 0) \equiv g(t, x')$. Moreover, there exists a constant N depending only on $\lambda, K, \kappa, d, \delta$ such that*

$$|u|_{1+\delta/2,2+\delta;Q} \le N\Big(|f|_{\delta/2,\delta;Q} + |g|_{1+\delta/2,2+\delta;\mathbb{R}^d}\Big). \tag{9.4.1}$$

We can obviously repeat the argument in Sec. 9.2 which yielded Theorem 9.2.3. Then we see that the only thing we need is the following result.

THEOREM 9.4.2. *Theorem 9.4.1 holds true if we replace equation (8.0.1) with equation (9.1.1).*

Proof. As in the proof of Theorem 9.1.2 we notice that uniqueness follows from the maximum principle. Also, we can continue f for negative x^d as $f(t, x', x^d) = f(t, x', 0)$, and we can repeat the corresponding argument from the proof of Theorem 9.1.2, again showing that we need to prove only the present theorem for $f \equiv 0$.

In this situation we define

$$p(t,x) = \begin{cases} \frac{1}{(4\pi t)^{d/2}} \frac{x^d}{t} e^{-\frac{1}{4t}|x|^2} & \text{for} \quad t > 0, x^d > 0\,, \\ 0 & \text{for} \quad t \le 0, x^d > 0\,, \end{cases} \qquad p_1(t,x) = p(t,x)e^{-t}.$$

We will use the fact that the function $p_1(t, x)$ is infinitely differentiable in (t, x) and for any $\varepsilon > 0$ any of its derivatives is bounded and integrable over the set $Q(\varepsilon) = (0, \varepsilon e_d) + Q$. Also,

$$\int_0^\infty \int_{\mathbb{R}^{d-1}} p(t, x', x^d)\, dx'dt = 1, \quad \Delta p - p_t = 0, \quad \Delta p_1 - p_{1t} - p_1 = 0, \tag{9.4.2}$$

$$\int_0^\infty \int_{\mathbb{R}^{d-1}} p(t, x', x^d)(t^{\delta/2} + |x'|^\delta)\, dx'dt = \gamma_1 |x^d|^\delta, \tag{9.4.3}$$

$$\int_0^\infty \int_{\mathbb{R}^{d-1}} |p_{x^d}(t, x', x^d)|(t^{\delta/2} + |x'|^\delta)\, dx'dt = \gamma_2 |x^d|^{\delta-1}, \tag{9.4.4}$$

where $\gamma_i = \gamma_i(d, \delta) < \infty$ and the last two equalities are obtained after an appropriate change of coordinates. Finally, for $x^d > 0$ let

$$u(t,x) = u(t, x', x^d) = p_1(\cdot, \cdot, x^d) * g(\cdot, \cdot)(t, x')$$

$$= \int_{-\infty}^\infty \int_{\mathbb{R}^{d-1}} p_1(t - s, x' - y', x^d) g(s, y')\, dy'ds$$

$$= \int_0^\infty \int_{\mathbb{R}^{d-1}} p(s, y', 1) e^{-s(x^d)^2} g(t - s(x^d)^2, x' - y'x^d)\, dy'ds,$$

where the last equality is obtained by changing variables. From what has been said above it follows that $u(t,x)$ is infinitely differentiable in Q and every derivative of $u(t,x)$ is bounded whenever x^d is bounded away from zero. In particular, $u \in C^{1+\delta/2, 2+\delta}(Q(\varepsilon))$ for any $\varepsilon > 0$. Moreover, from (9.4.2) it follows that $\Delta u - u_t - u = 0$ in Q. Furthermore,

$$\lim_{x^d \downarrow 0} u(t,x) = g(t,x') \int_0^\infty \int_{\mathbb{R}^{d-1}} p(s,y',1)\, dy'ds = g(t,x'). \tag{9.4.5}$$

Equations (9.4.2) also imply that $|u|_{0;Q} \le |g|_{0;\mathbb{R}^d}$, so that owing to interpolation inequalities we need to estimate only $[u]_{1+\delta/2,2+\delta;Q}$.

Let D be either D_iD_j with $i,j = 1,...,d-1$ or D_t. Obviously,

$$Du(t,x) = \int_0^\infty \int_{\mathbb{R}^{d-1}} p(s,y',1) e^{-s(x^d)^2} Dg(t - s(x^d)^2, x' - y'x^d)\, dy'ds,$$

$$\frac{|Du(t_1,x) - Du(t_2,x)|}{|t_1 - t_2|^{\delta/2}}$$

$$\le \int_0^\infty \int_{\mathbb{R}^{d-1}} p(s,y',1) e^{-s(x^d)^2} |t_1 - t_2|^{-\delta/2} \Big| Dg(t_1 - s(x^d)^2, x' - y'x^d)$$

$$- Dg(t_2 - s(x^d)^2, x' - y'x^d)\Big|\, dy'ds \le |g|_{1+\delta/2,2+\delta;\mathbb{R}^d}.$$

In the same way

$$\frac{|Du(t,x_1',x^d) - Du(t,x_2',x^d)|}{|x_1' - x_2'|^\delta} \le |g|_{1+\delta/2,2+\delta;\mathbb{R}^d}.$$

Now we are going to estimate Hölder's constants of Du with respect to x^d. The reader might find it instructive to compare the following argument with the one we have suggested in Exercise 9.3.1.

In the case $x_1^d + x_2^d \le 2|x_1^d - x_2^d|$ we notice that

$$\Big| Dg(t - s(x_1^d)^2, x' - y'x_1^d) - Dg(t - s(x_2^d)^2, x' - y'x_2^d)\Big|$$

$$\le [Dg]_{\delta/2,\delta}[s^{\delta/2}|(x_1^d)^2 - (x_2^d)^2|^{\delta/2} + |y'|^\delta |x_1^d - x_2^d|^\delta] \le N[Dg]_{\delta/2,\delta}(s^{\delta/2} + |y'|^\delta)|x_1^d - x_2^d|^\delta,$$

$$\frac{|Du(t,x',x_1^d) - Du(t,x',x_2^d)|}{|x_1^d - x_2^d|^\delta}$$

$$\le |Dg|_0 \int_0^\infty \int_{\mathbb{R}^{d-1}} p(s,y',1)|x_1^d - x_2^d|^{-\delta}|e^{-s(x_1^d)^2} - e^{-s(x_2^d)^2}|\, dy'ds$$

$$+ N[Dg]_{\delta/2,\delta;\mathbb{R}^d} \int_0^\infty \int_{\mathbb{R}^{d-1}} p(s,y',1)(s^{\delta/2} + |y'|^\delta)|\, dy'ds. \tag{9.4.6}$$

Since $|e^{-p^2} - e^{-q^2}| \leq N|p-q|^\delta$ if $p,q \geq 0$, by (9.4.3) we see that in the case $x_1^d + x_2^d \leq 2|x_1^d - x_2^d|$ ratio (9.4.6) is less than $N|g|_{1+\delta/2,2+\delta;\mathbb{R}^d}$.

If $x_1^d + x_2^d \geq 2|x_1^d - x_2^d|$, we use the fact that (see (9.4.2) and (9.4.4))

$$\frac{\partial}{\partial x^d} Du(t,x) = \int_{-\infty}^{\infty} \int_{\mathbb{R}^{d-1}} p_{1x^d}(t-s, x'-y', x^d) Dg(s,y')\, dy' ds$$

$$= \int_{-\infty}^{\infty} \int_{\mathbb{R}^{d-1}} p_{1x^d}(t-s, x'-y', x^d)[Dg(s,y') - Dg(t,x')]\, dy' ds,$$

$$|\frac{\partial}{\partial x^d} Du(t,x)|$$

$$\leq |g|_{1+\delta/2,2+\delta} \int_{-\infty}^{\infty} \int_{\mathbb{R}^{d-1}} |p_{1x^d}(t-s, x'-y', x^d)|(|t-s|^{\delta/2} + |x'-y'|^\delta)\, dy' ds$$

$$= N(x^d)^{\delta-1}|g|_{1+\delta/2,2+\delta;\mathbb{R}^d}.$$

We also apply the mean value theorem to the interval with the ends x_1^d, x_2^d and notice that $2\min(x_1^d, x_2^d) \geq |x_1^d - x_2^d|$. Then we see that the fraction in (9.4.6) is less than

$$N|g|_{1+\delta/2,2+\delta;\mathbb{R}^d}[\min(x_1^d, x_2^d)]^{\delta-1}|x_1^d - x_2^d|^{1-\delta} \leq N|g|_{1+\delta/2,2+\delta;\mathbb{R}^d}.$$

Thus, $[Du]_{\delta/2,\delta;Q} \leq N|g|_{1+\delta/2,2+\delta;\mathbb{R}^d}$. Moreover, from the equation we find that

$$[u_{x^dx^d}]_{\delta/2,\delta;Q} \leq \sum_{i,j=1}^{d-1} [u_{x^ix^j}]_{\delta/2,\delta;Q} + [u_t]_{\delta/2,\delta;Q} + [u]_{\delta/2,\delta;Q}.$$

By Corollary 9.3.3 we could get (9.4.1) if we knew that $u \in C^{1+\delta/2,2+\delta}(Q)$. But we know that $u(\cdot,\cdot,\cdot+\varepsilon) \in C^{1+\delta/2,2+\delta}(Q)$ for any $\varepsilon > 0$. Therefore, we can apply Corollary 9.3.3 to the shifted u. After this it remains only to make an obvious passage to the limit. The theorem is proved.

EXERCISE* 9.4.3. Prove (9.4.2) and (9.4.3), and also show that if $\delta = 1$, then the integrals in (9.4.3) diverge.

EXERCISE* 9.4.4 (cf. Exercises 8.9.3 and 9.2.4). Prove that there exists a constant N depending only on d, κ, δ, K such that for any $\lambda \geq 1$ and $u \in C^{1+\delta/2,2+\delta}(Q)$ we have

$$|u|_{1+\delta/2,2+\delta;Q} \leq N\Big([f]_{\delta/2,\delta;Q} + \lambda^{\delta/2}|f|_{0;Q} + [u(\cdot,0)]_{1+\delta/2,2+\delta;\mathbb{R}^d}$$

$$+\lambda^{1+\delta/2}|u(\cdot,0)|_{0;\mathbb{R}^d}\Big), \quad |u|_{0;Q} \leq \lambda^{-1}|f|_{0;Q} + |u(\cdot,0)|_{0;\mathbb{R}^d},$$

where $f = Lu - u_t - \lambda u$.

EXERCISE* 9.4.5. By using Exercises 9.4.4 and 8.8.2 prove that there exists a constant N depending only on d, κ, δ, K such that for any $\lambda \geq 1$ and $u \in C^{1+\delta/2,2+\delta}(Q)$ such that $u(\cdot, 0) \equiv 0$ we have

$$|u|_{\delta/2,\delta;Q} + |u_x|_{\delta/2,\delta;Q} \leq N\lambda^{(\delta-1)/2}|Lu - u_t - \lambda u|_{\delta/2,\delta;Q}.$$

9.5. Hints to exercises

9.1.3. By substituting $v(t,x) = u(Tt, \sqrt{T}\,x)$ reduce the whole thing to the case $T = 1$. In this case denote by w the solution from Theorem 9.1.2 corresponding to fe^{-t} and g and observe that the function we^t is the one we need.
9.1.4. By using dilations in (9.1.2) prove (9.1.5) for $Q = (0,T) \times \mathbb{R}^d$ and any T. Then let $\lambda \downarrow 0$.
9.1.5. See the proof of Theorem 8.7.2.
9.1.7. Apply the Fourier transform.
9.2.4. See the hint to Exercise 8.9.3.
9.4.4. See the hint to Exercise 8.9.3.

CHAPTER 10

Parabolic Equations in Domains

In this chapter, as in Chapter 9, we consider elliptic second-order operators L with real coefficients as in (8.0.2) (in particular $c \leq 0$). We always assume that L is nondegenerate with constant of ellipticity $\kappa > 0$. Also we assume that $|a, b, c|_{\delta/2,\delta} \leq K$, where $\delta \in (0, 1)$ and K are fixed.

10.1. Preliminaries

Let $T \in (-\infty, \infty]$ and Ω be a (bounded) domain of class $C^{2+\delta}$ in $\mathbb{R}^d$ (see Definition 6.1.6). Let $Q = (-\infty, T) \times \Omega$ and define

$$C_0^{1+\delta/2,2+\delta}(Q) = C^{1+\delta/2,2+\delta}(\bar{Q}) \cap \{u : u = 0 \text{ on } \partial' Q\}.$$

In Sec. 6.2 we described how elliptic operators transform upon straightening the boundary of Ω. We will do the same transformation, fixing a point $x_0 \in \partial\Omega$ and using the same function ψ, which, by the way, is independent of t. The latter observation shows that in our new situation we, of course, have the corresponding counterparts of Lemmas 6.1.8 and 6.2.1. As in Sec. 6.2 define, for $y = \psi(x)$,

$$\tilde{a}^{kl}(t, y) = a^{ij}(t, x)\psi^k_{x^i}(x)\psi^l_{x^j}(x), \quad \tilde{b}^k(t, y) = a^{ij}(t, x)\psi^k_{x^i x^j}(x) + b^i(t, x)\psi^k_{x^i}(x),$$

$$\tilde{c}(t, y) = c(t, x), \quad \tilde{\eta}(y) = \eta(x - x_0),$$

$$\tilde{L}w(t, y) := \tilde{a}^{kl}(t, y)w_{y^k y^l}(t, y) + \tilde{b}^k(t, y)w_{y^k}(t, y) + \tilde{c}(t, y)w(t, y),$$

$$\bar{L}(t, y) = \tilde{\eta}(y)\tilde{L}(t, y) + (1 - \tilde{\eta}(y))\Delta.$$

Next, let $E = \{(t, x', x^d) : t < T, x' \in \mathbb{R}^{d-1}, x^d > 0\}$. By Theorem 9.4.1 for any function $f = f(t, y)$ of class $C^{\delta/2,\delta}(E)$ and any $\lambda > 0$ there exists a unique solution $u \in C^{1+\delta/2,2+\delta}(E)$ of the equation $\bar{L}u - u_t - \lambda u + f = 0$ in E satisfying the boundary condition $u(t, x', 0) \equiv 0$. We call this solution $\bar{\mathcal{R}}_\lambda f$. By Exercises 9.4.4 and 9.4.5 for $\lambda \geq 1$ we have

$$|\bar{\mathcal{R}}_\lambda f|_{1+\delta/2,2+\delta;E} \leq N\Big([f]_{\delta/2,\delta;E} + \lambda^{\delta/2}|f|_{0;E}\Big),$$

$$|\bar{\mathcal{R}}_\lambda f|_{\delta/2,\delta;E} + |(\bar{\mathcal{R}}_\lambda f)_x|_{\delta/2,\delta;E} \leq N\lambda^{(\delta-1)/2}|f|_{\delta/2,\delta;E}, \tag{10.1.1}$$

the constant N depending only on d, κ, δ, K.

Further, as in Sec. 6.2 we define the operators

$$\Psi : w = w(t, y) \to \Psi w(t, x) = w(t, \psi(x)),$$

$$\Psi^{-1} : v = v(t, x) \to \Psi^{-1}v(t, y) = v(t, \psi^{-1}(y)),$$

$$S_\lambda^{x_0}: f = f(t,x) \to S_\lambda^{x_0} f(t,x) = \Psi \bar{\mathcal{R}}_\lambda \Psi^{-1}[\eta(\cdot - x_0) f](t,x),$$

where by $\Psi^{-1}[\eta(\cdot - x_0)f](t,y)$ we certainly mean zero for $y \notin D_+$ and the function $\eta(\psi^{-1}(y) - x_0) f(t, \psi^{-1}(y))$ for $y \in D_+$. By using (10.1.1) and repeating the proof of Theorem 6.2.2 we get the following result in which

$$C_\rho(x_0) := (-\infty, T) \times B_\rho(x_0).$$

THEOREM 10.1.1. *(i) If $v \in C_0^{1+\delta/2,2+\delta}(Q)$ and $v = 0$ outside $C_{\rho_0/2}(x_0) \cap Q$, then in $C_{\rho_0/2}(x_0) \cap Q$ for any $\lambda \geq 1$ we have*

$$v = S_\lambda^{x_0}(\lambda v + v_t - Lv).$$

(ii) There is a constant N depending only on $\kappa, K_0, \rho_0, \delta, K, d$ such that for $\lambda \geq 1$ and $f \in C^{\delta/2,\delta}(Q)$ we have

$$|S_\lambda^{x_0} f|_{1+\delta/2,2+\delta;C_{\rho_0}(x_0)\cap Q} \leq N\Big([f]_{\delta/2,\delta;C_{\rho_0}(x_0)\cap Q} + \lambda^{\delta/2}|f|_{0;C_{\rho_0}(x_0)\cap Q}\Big), \quad (10.1.2)$$

$$|S_\lambda^{x_0} f|_{\delta/2,\delta;C_{\rho_0}(x_0)\cap Q} + |(S_\lambda^{x_0} f)_x|_{\delta/2,\delta;C_{\rho_0}(x_0)\cap Q} \leq N\lambda^{(\delta-1)/2}|f|_{\delta/2,\delta;C_{\rho_0}(x_0)\cap Q}.$$

REMARK 10.1.2. Assertion (i) of Theorem 10.1.1 says that if we want to solve the equation $\lambda u + u_t - Lu = f$ in $C_0^{1+\delta/2,2+\delta}(Q)$ and we know in advance that the solution vanishes outside $C_{\rho_0/2}(x_0) \cap Q$, then the solution in $C_{\rho_0/2}(x_0) \cap Q$ is given by $S_\lambda^{x_0} f$.

EXERCISE* 10.1.3. Prove that if $f \in C^{\delta/2,\delta}(C_{\rho_0}(x_0) \cap Q)$, then

$$S_\lambda^{x_0} f \in C^{1+\delta/2,2+\delta}(C_{\rho_0}(x_0) \cap Q), \quad S_\lambda^{x_0} f = 0 \quad \text{on} \quad C_{\rho_0}(x_0) \cap \partial' Q$$

and $(\lambda + \partial/\partial t - L) S_\lambda^{x_0} f = f$ in $C_{\rho_0/2}(x_0) \cap Q$.

10.2. Some applications of partitions of unity

We keep the notation from Sec. 10.1. In particular, $Q = (-\infty, T) \times \Omega$. Take $n, x_1, ..., x_n, \zeta^i$ from Sec. 6.3. In the same way as Lemma 6.3.1 one proves the following lemma.

LEMMA 10.2.1. *There exists a constant $N = N(d, \delta, K_0, \rho_0, d_\Omega)$ such that for any $v \in C^{1+\delta/2,2+\delta}(Q)$ we have*

$$|v|_{\delta/2,\delta;Q} \leq N\Big(|v|_{1+\delta/2,2+\delta;Q}^{\delta/(2+\delta)} |v|_{0;Q}^{2/(2+\delta)} + |v|_{0;Q}\Big),$$

$$|v_x|_{\delta/2,\delta;Q} \leq N\Big(|v|_{1+\delta/2,2+\delta;Q}^{(1+\delta)/(2+\delta)} |v|_{0;Q}^{1/(2+\delta)} + |v|_{0;Q}\Big),$$

$$|v_{xx}|_{0;Q} \leq N\Big(|v|_{1+\delta/2,2+\delta;Q}^{2/(2+\delta)} |v|_{0;Q}^{\delta/(2+\delta)} + |v|_{0;Q}\Big),$$

$$|v_t|_{0;Q} \leq N\Big(|v|_{1+\delta/2,2+\delta;Q}^{2/(2+\delta)} |v|_{0;Q}^{\delta/(2+\delta)} + |v|_{0;Q}\Big).$$

In particular, for any $\varepsilon > 0$ there is a constant $N(\varepsilon)$ depending only on ε, d, δ, K_0, ρ_0, d_Ω such that

$$|v_{xx}|_{0;Q} + |v_t|_{0;Q} + |v_x|_{\delta/2,\delta;Q} + |v|_{\delta/2,\delta;Q} \le \varepsilon |v|_{1+\delta/2,2+\delta;Q} + N(\varepsilon)|v|_{0;Q}.$$

In the same way as in Sec. 6.3, Lemma 10.2.1 and previous results along with Theorem 8.1.8 lead us to the following a priori estimate (see the proof of Theorem 6.3.2).

THEOREM 10.2.2. *There is a constant N depending only on d_Ω, δ, d, κ, K, K_0, ρ_0 such that for any $u, g \in C^{1+\delta/2,2+\delta}(Q)$ satisfying $u = g$ on $\partial' Q$, we have*

$$|u|_{1+\delta/2,2+\delta;Q} \le N\Big(|Lu - u_t|_{\delta/2,\delta;Q} + |g|_{1+\delta/2,2+\delta;Q}\Big). \tag{10.2.1}$$

REMARK 10.2.3. If we knew that the equation $\Delta u - u_t - u = f$ is solvable in $C_0^{1+\delta/2,2+\delta}(Q)$, then by the method of continuity we could get the solvability of the equation $Lu - u_t = f$ on the sole basis of estimate (10.2.1).

Partitions of unity allow us to define the boundary spaces $C^{1+\delta/2,2+\delta}(\partial' Q)$ and to identify them with the spaces of functions which admit extensions of class $C^{1+\delta/2,2+\delta}$ inside Q. Below we denote by ψ_i, D^i the mapping $B_{\rho_0}(x_i) \to D^i$ and the domain $D^i \subset \mathbb{R}^d$ corresponding to x_i in Definition 6.1.6.

DEFINITION 10.2.4. Let g be a function defined on $\partial' Q$. We will write $g \in C^{1+\delta/2,2+\delta}(\partial' Q)$ if for any $i = 1, ..., n$ for the functions g_i defined in

$$\partial' E = \{(t, y) : t < T, y \in \mathbb{R}^d, y^d = 0\}$$

by the formula $g_i(t, y) := \Psi_i^{-1}(g\zeta^i)(t, y) = g(t, \psi_i^{-1}(y))\zeta^i(\psi_i^{-1}(y))$ we have $g_i \in C^{1+\delta/2,2+\delta}(\partial' E)$. We also write

$$|g|_{1+\delta/2,2+\delta;\partial' Q} = \sum_{i=1}^{n} |g_i|_{1+\delta/2,2+\delta;\partial' E}.$$

THEOREM 10.2.5. *A function $g \in C^{1+\delta/2,2+\delta}(\partial' Q)$ if and only if there exists a function $\tilde{g} \in C^{1+\delta/2,2+\delta}(Q)$ such that $\tilde{g} = g$ on $\partial' Q$. For any such extension $\tilde{g}$ we have*

$$|g|_{1+\delta/2,2+\delta;\partial' Q} \le N|\tilde{g}|_{1+\delta/2,2+\delta;Q}, \tag{10.2.2}$$

where the constant N depends only on d, δ, K_0, ρ_0. With the same kind of constant N for any $g \in C^{1+\delta/2,2+\delta}(\partial' Q)$ there exists $h \in C^{1+\delta/2,2+\delta}(Q)$ such that $h = g$ on $\partial' Q$ and

$$|h|_{1+\delta/2,2+\delta;Q} \le N|g|_{1+\delta/2,2+\delta;\partial' Q}. \tag{10.2.3}$$

Proof. If g is the restriction on $\partial' Q$ of a function $\tilde{g} \in C^{1+\delta/2,2+\delta}(Q)$, then

$$|g_i|_{1+\delta/2,2+\delta;\partial' E} = |\Psi_i^{-1}(\tilde{g}\zeta^i)|_{1+\delta/2,2+\delta;\partial' E} \le |\Psi_i^{-1}(\tilde{g}\zeta^i)|_{1+\delta/2,2+\delta;(-\infty,T)\times D_+^i}$$
$$\le N|\tilde{g}\zeta^i|_{1+\delta/2,2+\delta;C_{\rho_0}(x_i)\cap Q} \le N|\tilde{g}\zeta^i|_{1+\delta/2,2+\delta;Q} \le N|\tilde{g}|_{1+\delta/2,2+\delta;Q}.$$

This proves that $g \in C^{1+\delta/2,2+\delta}(\partial' Q)$ and that (10.2.2) holds.

On the other hand, take $g \in C^{1+\delta/2,2+\delta}(\partial' Q)$ and for $y = (y', y^d) \in \mathbb{R}^d$, $t < T$, $x \in \Omega$ define

$$\bar{g}_i(t, y) = g_i(t, y', 0), \quad h_i(t, x) = \bar{g}_i(t, \psi_i(x))\zeta^i(x), \quad h(t, x) = \sum_{i=1}^n h_i(t, x).$$

Then for $x \in (\partial\Omega) \cap B_{\rho_0}(x_i)$ and $y = \psi_i(x)$ we have $y^d = 0$ and

$$h_i(t, x) = \bar{g}_i(t, y)\zeta^i(x) = g_i(t, y', 0)\zeta^i(x) = g(t, x)[\zeta^i(x)]^2.$$

The last equalities are also true for $x \in (\partial\Omega) \setminus B_{\rho_0}(x_i)$, since then $\zeta^i(x) = 0$. It follows that for $x \in \partial\Omega$

$$h(t, x) = \sum_{i=1}^n g(t, x)[\zeta^i(x)]^2 = g(t, x).$$

Finally,

$$|h_i|_{1+\delta/2,2+\delta;Q} \leq N|h_i|_{1+\delta/2,2+\delta;C_{\rho_0}(x_i)\cap Q} \leq N|\Psi_i^{-1} h_i|_{1+\delta/2,2+\delta;(-\infty,T)\times D_+^i}$$

$$= N|\bar{g}_i \Psi_i^{-1} \zeta_i|_{1+\delta/2,2+\delta;(-\infty,T)\times D_+^i} \leq N|\bar{g}_i|_{1+\delta/2,2+\delta;E} = N|g_i|_{1+\delta/2,2+\delta;\partial' E},$$

which proves (10.2.3). The theorem is proved.

EXERCISE 10.2.6. By imitating the proof of Theorem 8.11.1 and using (10.2.1) prove that if $z \in \mathbb{R}^{d+1}$ and we are given a function $u \in C^{1+\delta/2,2+\delta}(Q_{3R}(z) \cap Q)$ such that $u = 0$ on $Q_{3R}(z) \cap \partial' Q$, then

$$|u|_{1+\delta/2,2+\delta;Q_R(z)\cap Q} \leq N\Big(|Lu - u_t|_{\delta/2,\delta;Q_{2R}(z)\cap Q} + |u|_{0;Q_{2R}(z)\cap Q}\Big),$$

where N depends only on $R, d_\Omega, \delta, d, \kappa, K, K_0, \rho_0$.

10.3. First boundary-value problem in infinite smooth cylinders

Now we are ready to consider the first boundary-value problem for the domain $Q = (-\infty, T) \times \Omega$ as in two previous sections. Denote $S_\lambda^i = S_\lambda^{x_i}$, $i = 1, ..., n$ and $S_\lambda^0 = \mathcal{R}_\lambda$ where we take the last operator from Remark 8.10.3.

As in Sec. 6.4 the main idea of proving solvability of the equation $Lu - u_t + f = 0$ in $C_0^{1+\delta/2,2+\delta}(Q)$ is to reduce the equation to a kind of integral equation related to the notion of regularizer.

The following fact is proved in the same way as Lemma 6.4.1.

LEMMA 10.3.1. *Let $\lambda \geq 1$ and $u \in C_0^{1+\delta/2,2+\delta}(Q)$ and $\lambda u + u_t - Lu = f$. Then in Q*

$$u = \sum_{i \leq n} \zeta^i S_\lambda^i(\zeta^i f - L^i u), \tag{10.3.1}$$

where

$$L^k u := u(a^{ij}\zeta_{x^i x^j}^k + b^i \zeta_{x^i}^k) + 2a^{ij}\zeta_{x^i}^k u_{x^j} \quad (= L(\zeta^k u) - \zeta^k Lu).$$

This lemma suggests that solutions of the equation $\lambda u + u_t - Lu = f$ should be looked for in the set of solutions of (10.3.1). To show that it is indeed enough to solve (10.3.1), we introduce the Banach space $D^{1+\delta}(Q)$ (cf. Remark 8.8.7) as the set of all functions $u(t,x)$ which are continuously differentiable with respect to x and have finite norm

$$||u||_{1+\delta;Q} := |u|_{\delta/2,\delta;Q} + |u_x|_{\delta/2,\delta;Q}.$$

LEMMA 10.3.2. *If $\lambda \geq 1$ and $f \in C^{\delta/2,\delta}(Q)$, and $u \in D^{1+\delta}(Q)$ is a solution of (10.3.1), then $u \in C_0^{1+\delta/2,2+\delta}(Q)$. Furthermore, there exists*

$$\lambda_0 = \lambda_0(d, K, \kappa, \delta, K_0, \rho_0, d_\Omega) \geq 1$$

such that if, in addition, $\lambda \geq \lambda_0$, then $\lambda u + u_t - Lu = f$ in Q.

Proof. The inclusion $u \in C_0^{1+\delta/2,2+\delta}(Q)$ follows from the fact that $L^i u \in C^{\delta/2,\delta}(Q)$, whence $\zeta^i S^i_\lambda(\zeta^i f - L^i u) \in C_0^{1+\delta/2,2+\delta}(Q)$ for any i. To prove that if λ is large enough, then u satisfies the equation $\lambda u + u_t - Lu = f$ in Q, we proceed as in the proof of Lemma 6.5.1, and we see that we need to prove only that the operator

$$T_\lambda : h \to T_\lambda h := \sum_{i=0}^{n} L^i S^i_\lambda(\zeta^i h)$$

is a contraction in $C^{\delta/2,\delta}(Q)$.

By Theorem 10.1.1 we have

$$|L^i S^i_\lambda(\zeta^i h)|_{\delta/2,\delta;Q} \leq N||S^i_\lambda(\zeta^i h)||_{1+\delta;C_{\rho_0}(x_i)\cap Q}$$

$$\leq N\lambda^{(\delta-1)/2}|\zeta^i h|_{\delta/2,\delta;C_{\rho_0}(x_i)\cap Q} \leq N\lambda^{(\delta-1)/2}|h|_{\delta/2,\delta;Q}.$$

By using Exercise 8.9.4 we obtain a similar inequality for $i = 0$. Since $\lambda^{(\delta-1)/2}$ can be chosen as small as we like, the lemma is proved.

THEOREM 10.3.3. *For any $f \in C^{\delta/2,\delta}(Q)$ and $g \in C^{1+\delta/2,2+\delta}(Q)$ there exists a unique function $u \in C^{1+\delta/2,2+\delta}(\bar{Q})$ satisfying the equation $Lu - u_t = f$ in Q and equal g on $\partial' Q$.*

Proof. Uniqueness follows from Theorem 8.1.8. As far as existence is concerned, by considering $u - g$ instead of u, one reduces the general case to the case $g = 0$. Then, consider the equation $Lu - u_t - \lambda u = f$ in $C_0^{1+\delta/2,2+\delta}(Q)$ for λ large enough. Lemma 10.3.2 shows that to prove the solvability of this equation it suffices to prove that there exists a function $u \in D^{1+\delta}(Q)$ which solves (10.3.1).

In the same way as in the end of the proof of Lemma 10.3.2 one shows that the operator

$$u \to \sum_{i=0}^{n} \zeta^i S^i_\lambda(L^i u)$$

is a contraction in $D^{1+\delta}(Q)$ if λ is large enough. Therefore we have the solvability of $Lu - u_t - \lambda u = f$ indeed. To finish the proof it remains only to fix an appropriate λ and apply the method of continuity to the family of equations $Lu - u_t - \mu\lambda u = f$,

$\mu \in [0, 1]$, remembering that Theorem 10.2.2 provides us with the necessary a priori estimate. The theorem is proved.

REMARK 10.3.4. If $g(t, x) = f(t, x) = 0$ for all $t \leq 0$, then by Theorem 8.1.8 we have $u(t, x) = 0$ for $t \leq 0$. This fact will be used in the sequel to solve parabolic equations in $(0, \infty) \times \Omega$.

REMARK 10.3.5. As in the elliptic case the above proof can easily be adjusted to parabolic equations when Ω is a smooth manifold with or without boundary.

REMARK 10.3.6. We have constructed solutions only in straight cylinders. Actually the same method based on the notion of regularizer applies when the shape of sections of Q by planes parallel to $\mathbb{R}^d$ varies. Nevertheless, in a neighborhood of every point on $\partial' Q$ the domain Q should be close to a part of a spacelike half space.

EXERCISE 10.3.7. Prove that for any $R > 0$ if $u \in C^{1+\delta/2, 2+\delta}(C_{3R} \cap Q)$ and $u = 0$ on $C_{3R} \cap \partial Q$, then

$$|u|_{1+\delta/2,2+\delta;C_R \cap Q} \leq N\Big(|Lu - u_t|_{\delta/2,\delta;C_{2R}\cap Q} + |u|_{0;C_{2R}\cap Q}\Big),$$

where N is independent of u.

10.4. Mixed problem

Let $Q = (0, \infty) \times \Omega$, where the domain $\Omega \subset \mathbb{R}^d$ and Ω is of class $C^{2+\delta}$. By the mixed problem we mean the following:

$$Lu - u_t = f \quad \text{in} \quad Q, \quad u = g \quad \text{on} \quad \partial_x Q, \quad u(0, x) = h(x) \quad \text{for} \quad x \in \bar{\Omega}. \quad (10.4.1)$$

This amounts to solving the degenerate elliptic equation $Lu - u_t = f$ in the domain Q whose boundary is not smooth. In Exercise 7.5.2 we saw what happens even to the best elliptic equation when the domain is not smooth.

In (10.4.1) we have the same kind of situation. If we want that $u \in C^{1+\delta,2+\delta}(Q)$, then u will be extendible to $\bar{Q}$ within the same class of functions. In particular, g and h should be boundary values of a continuous function and should satisfy *the zeroth consistency condition*:

$$g(0, x) = h(x) \quad \text{for} \quad x \in \partial\Omega. \quad (10.4.2)$$

Furthermore, for $x \in \partial\Omega$ we have $u_t(t, x) = g_t(t, x)$ and $u_t(0, x) = g_t(0, x)$. Also $L(0, x)u(0, x) = L(0, x)h(x)$ for $x \in \bar{\Omega}$, so that for $x \in \partial\Omega$ we should have $f(0, x) = L(0, x)u(0, x) - u_t(0, x) = L(0, x)h(x) - g_t(0, x)$. We come to *the first consistency condition*:

$$f(0, x) = L(0, x)h(x) - g_t(0, x) \quad \text{for} \quad x \in \partial\Omega. \quad (10.4.3)$$

In the following theorem we denote $\mathbb{R}^{d+1}_+ = (0, \infty) \times \mathbb{R}^d$, and we assume that some data are given and belong to appropriate Hölder classes on sets which are larger than is needed. Actually we need them to be given only in Q or on $\partial' Q$. In connection with this we mention extension theorems (see, for instance, [**6**]) which assert that such extensions exist always, so that the hypotheses of the theorem are not restrictive at all.

THEOREM 10.4.1. *Let $f \in C^{\delta/2,\delta}(\mathbb{R}^{d+1}_+)$, $g \in C^{1+\delta/2,2+\delta}(Q)$, $h \in C^{2+\delta}(\mathbb{R}^d)$. Assume that (10.4.2) and (10.4.3) are satisfied. Then there exists a unique function $u \in C^{1+\delta/2,2+\delta}(Q)$ satisfying (10.4.1). In addition, there exists a constant N depending only on $d, \kappa, \delta, K, \rho_0, K_0, d_\Omega$ such that*

$$|u|_{1+\delta/2,2+\delta;Q} \le N\Big(|f|_{\delta/2,\delta;\mathbb{R}^{d+1}_+} + |g|_{1+\delta/2,2+\delta;Q} + |h|_{2+\delta;\mathbb{R}^d}\Big). \tag{10.4.4}$$

Proof. Uniqueness follows from Corollary 8.1.5. In the proof of existence first assume that $c \le c_0$, where c_0 is a strictly negative constant, perhaps depending on d, κ, δ, K, ρ_0, K_0, d_Ω. Let v be a $C^{1+\delta/2,2+\delta}(\mathbb{R}^{d+1}_+)$–solution of the Cauchy problem:

$$Lv - v_t = f \quad \text{in} \quad \mathbb{R}^{d+1}_+, \quad v(0,x) = h(x) \quad \text{on} \quad \mathbb{R}^d$$

(see Theorem 9.2.3). To construct the function u in question, we now have to add to v a function w solving the following problem:

$$Lw - w_t = 0 \quad \text{in} \quad Q, \quad w(0,x) = 0 \quad \text{in} \quad \bar{\Omega}, \quad w = g - v =: g_1 \quad \text{on} \quad \partial_x Q. \tag{10.4.5}$$

From (10.4.2) and (10.4.3) for $x \in \partial\Omega$ we have that

$$g_1(0,x) = g(0,x) - v(0,x) = g(0,x) - h(x) = 0,$$

$$g_{1t}(0,x) = g_t(0,x) - v_t(0,x) = g_t(0,x) - (Lv)(0,x) + f(0,x)$$

$$= g_t(0,x) - L(0,x)h(x) + f(0,x) = 0.$$

These relations imply easily that if we set $g_1(t,x) = 0$ for $t < 0$, then $g_1 \in C^{1+\delta/2,2+\delta}(\mathbb{R} \times \partial\Omega)$. By Theorems 10.2.5 and 10.3.3 there is a $C^{1+\delta/2,2+\delta}(\mathbb{R} \times \Omega)$ function w such that

$$Lw - w_t = 0 \quad \text{in} \quad \mathbb{R} \times \Omega, \quad w = g_1 \quad \text{on} \quad \mathbb{R} \times \partial\Omega.$$

By Theorem 8.1.8 and since $w(t,x) = 0$ on $(-\infty, 0] \times \partial\Omega$, we have $w(t,x) = 0$ for $t \le 0$, so that w solves (10.4.5). To get (10.4.4) for $u := v + w$ it remains only to use the estimates contained in Theorems 9.2.3, 10.2.5 and 10.2.2.

Finally, to drop the extra condition $c \le c_0$, without loss of generality we assume that $0 \in \Omega$, and we take the function v_0 from Lemma 6.1.1 corresponding to $R = 2d_\Omega$. Then we introduce the functions $u' = u/v_0$, $(f', g', h') = (f, g, h)/v_0$, and the operator $L'w = v_0^{-1}L(v_0 w)$ and call a', b', c' its coefficients. Owing to the choice of v_0, we have $c' \le -1/v_0$. Also for these primed objects the consistency conditions are satisfied. Therefore we can solve (10.4.1) with L', f', g', h' instead of L, f, g, h. If u' is the solution, then $u = v_0 u'$ solves (10.4.1). The estimate for u is also easily derived from the estimate for u'. This proves the theorem.

EXERCISE 10.4.2. Take $\rho > 0$ and define $Q^{j\rho} = \{z \in Q : \operatorname{dist}(z, \partial_{tx} Q) > j\rho\}$, $j = 1, 2$, and $\Omega^\rho = \{x : (0,x) \in \bar{Q}^\rho\}$. By applying the results of Exercises 9.2.5 and 10.2.6 prove that under the conditions of Theorem 10.4.1

$$|u|_{1+\delta/2,2+\delta;Q^{2\rho}} \le N\Big(|f|_{\delta/2,\delta;Q^\rho} + |g|_{1+\delta/2,2+\delta;Q^\rho} + |h|_{2+\delta;\Omega^\rho} + |u|_{0;Q^\rho}\Big),$$

the constant N depending only on $\rho, d, \kappa, \delta, K, \rho_0, K_0, d_\Omega$.

EXERCISE 10.4.3. Under the conditions of Theorem 10.4.1 assume that for $t \geq 1$ we have $g(t,x) = 0$ and $f(t,x) = f(x)$. Also assume that the coefficients of L are independent of t. Prove that $u(t,x) \to v(x)$ as $t \to \infty$, where v is a $C_0^{2+\delta}(\Omega)$ solution of the equation $Lv(x) = f(x)$.

10.5. Convergence of numerical approximations

Take a bounded domain $\Omega \in C^{2+\delta}$ and a number $T \in (0,\infty)$. Denote $Q = (0,T) \times Q$. Let $l(h)$ be a function on $(0,1]$ such that $l(h) > 0$ and $l(h) \to 0$ as $h \downarrow 0$. For $h \in (0,1]$ define

$$Z_h^{d+1} = \{(t,x) : t = l(h)k, x = h\sum_{i=1}^{d} e_i n_i, k = 0,1,2,..., n_i = 0,\pm 1,\pm 2,...\}$$

the $(l(h),h)$–grid on $\mathbb{R}_+^{d+1}$. Also let $Q(h) = Q \cap Z_h^{d+1}$ and denote by $Q^o(h)$ the "interior" of $Q(h)$ that is the set of all points (t,x) in $Q(h)$ such that both dist $(x,\partial\Omega) \geq h$ and $t \geq l(h)$. The set $\partial' Q(h) = Q(h) \setminus Q^o(h)$ is interpreted as the discrete parabolic boundary of $Q(h)$.

Assume that for any $h \in (0,1], z \in Q^o(h), z_1 \in Q(h)$ we are given some numbers $p_h(z,z_1)$ and denote

$$\mathcal{L}_h u(z) = \sum_{z_1 \in Q(h)} p_h(z,z_1)u(z_1).$$

We want the operators $\mathcal{L}_h$ to "behave" like $L - \partial/\partial t$ and to approximate $L - \partial/\partial t$ as $h \downarrow 0$. To this end we make the following assumptions.

ASSUMPTION 10.5.1 (maximum principle). If u is a function defined on $Q(h)$ and for a point $z_0 \in Q^o(h)$ we have $u(z_0) = \max_{\Omega_h} u(z) > 0$, then $\mathcal{L}_h u(z_0) \leq 0$.

ASSUMPTION 10.5.2. The operators $\mathcal{L}_h$ approximate $L - \partial/\partial t$. More precisely, for any $u \in C^{1+\delta/2,2+\delta}(Q)$ and any $z \in Q^o(h)$ we have

$$|Lu(z) - u_t(z) - \mathcal{L}_h u(z)| \leq Kh^\delta |u|_{1+\delta/2,2+\delta;Q}.$$

One proves the following lemma by repeating the proof of Lemma 6.7.1.

LEMMA 10.5.3. *There is a constant $h_0 > 0$ depending only on κ, K, δ, d and the diameter of Ω such that for $h \in (0,h_0]$ for any bounded functions f,g the system of linear equations*

$$\mathcal{L}_h u(z) = f(z) \quad \forall z \in Q^o(h), \quad u(z) = g(z) \quad \forall z \in \partial' Q(h) \tag{10.5.1}$$

has a unique solution $u_h(z), z \in Q(h)$. In addition

$$\max_{Q(h)}(u_h(z))_+ \leq N \max_{Q^o(h)} f_+(z) + \max_{\partial' Q(h)} g_+(z),$$

$$\max_{Q(h)}(u_h(z))_- \leq N \max_{Q^o(h)} f_-(z) + \max_{\partial' Q(h)} g_-(z),$$

$$\max_{Q(h)} |u_h(z)| \leq N \max_{Q^o(h)} |f(z)| + \max_{\partial' Q(h)} |g(z)|,$$

where the constant N depends only on κ, K, d and the diameter of Ω.

THEOREM 10.5.4. *Let $f \in C^{\delta/2,\delta}(\mathbb{R}^{d+1}_+), g \in C^{1+\delta/2,2+\delta}(\mathbb{R}^{d+1}_+)$. In Theorem 10.4.1 take $h(x) = g(0,x)$ and assume that its hypotheses are satisfied. Let $u \in C^{1+\delta/2,2+\delta}(Q)$ be the solution of (10.4.1). Furthermore, take $h \in (0, h_0]$ and denote by u_h the corresponding solution of (10.5.1). Then*

$$|u - u_h|_{0;Q(h)} \leq Nh^\delta \Big(|f|_{\delta/2,\delta;\mathbb{R}^{d+1}_+} + |g|_{1+\delta/2,2+\delta;\mathbb{R}^{d+1}_+} \Big),$$

where the constant N depends only on d, K, δ, κ, the diameter of Ω and the constants ρ_0, K_0 from Definition 6.1.6.

To prove this theorem it suffices to follow the proof of Theorem 6.7.2 with obvious modifications.

Let us show how the operators $\mathcal{L}_h$ can be constructed in a particular case in which

$$L = \sum_{i=1}^{d} a^{ii}(t,x) D_i^2.$$

Let $l(h) = \varepsilon h^2$, where $\varepsilon^{-1} \geq 2 \sup_z \operatorname{tr} a(z)$, and for $(t,x) \in Q^o(h)$ denote

$$\mathcal{L}_h u(t,x) = -\frac{u(t,x) - u(t-\varepsilon h^2, x)}{\varepsilon h^2}$$

$$+ \sum_{i=1}^{d} a^{ii}(t - \varepsilon h^2, x) \frac{u(t-\varepsilon h^2, x + he_i) - 2u(t - \varepsilon h^2, x) + u(t - \varepsilon h^2, x - he_i)}{h^2}.$$

Owing to Taylor's formula and the assumption $a \in C^{\delta/2,\delta}(\mathbb{R}^{d+1})$, one easily checks that Assumption 10.5.2 is satisfied. To check Assumption 10.5.1 let $z_0 = (t_0, x_0) \in Q^o(h)$ and $u(t_0,x_0) = M = \max_{Q(h)} u(z) > 0$. Then

$$h^2 \mathcal{L}_h u(t_0, x_0) = -M\varepsilon^{-1} + u(t - \varepsilon h^2, x_0)[\varepsilon^{-1} - 2\sum_i a^{ii}(t_0 - \varepsilon h^2, x_0)]$$

$$+ \sum_i a^{ii}(t - \varepsilon h^2, x_0)[u(t-\varepsilon h^2, x_0 + he_i) + u(t - \varepsilon h^2, x_0 - he_i)]$$

$$\leq -M\varepsilon^{-1} + M[\varepsilon^{-1} - 2\sum_i a^{ii}(t_0 - \varepsilon h^2, x_0)] + 2M \sum_i a^{ii}(t_0 - \varepsilon h^2, x_0) = 0,$$

which means that Assumption 10.5.1 is satisfied too.

Thus, by solving (10.5.1) with this $\mathcal{L}_h$ we can approximate solutions of the mixed problem. Observe that the first equation in (10.5.1) is the following

$$u_h(t,x) = -\varepsilon h^2 f(t,x) + u_h(t',x)$$

$$+ \varepsilon \sum_i a^{ii}(t',x)[u_h(t', x + he_i) - 2u_h(t',x) + u_h(t', x - he_i)], \quad (10.5.2)$$

where $t' := t - \varepsilon h^2$. It is seen that one can compute $u_h(t,x)$ on $Q(h)$ consecutively, starting from $u(0,x)$ which is given, then finding $u(\varepsilon h^2, x)$ from the explicit formula (10.5.2), then finding $u(2\varepsilon h^2, x)$ again from (10.5.2), and so on. The method of approximation based on the above $\mathcal{L}_h$ is called an *explicit* method.

Under the hypotheses of Exercise 10.4.3 when T is large, $u(T,x)$ is close to a solution of an elliptic equation; thus, we get one more method of approximating solutions of elliptic equations. It turns out that actually this method essentially coincides with the following method of solving (6.7.1): rewrite (6.7.1) as

$$u(x) = \sum_{y \neq x} \frac{p_h(x,y)}{|p_h(x,x)|} u(y) + \frac{1}{|p_h(x,x)|} f(x),$$

(observe that by Exercise 6.6.4 we have $p_h(x,x) \leq 0$), then take any $u(0,x)$, find $u(1,x)$ by substituting $u = u(0,\cdot)$ into the right-hand side, and so on. In other words, one defines

$$u(n,x) = \sum_{y \neq x} \frac{p_h(x,y)}{|p_h(x,x)|} u(n-1,y) + \frac{1}{|p_h(x,x)|} f(x),$$

or equivalently

$$u(n,x) = u(n-1,x) + \frac{1}{|p_h(x,x)|} L_h u(n-1,x) + \frac{1}{|p_h(x,x)|} f(x).$$

There also exist implicit methods. For example, take any constant $\varepsilon > 0$ and $l(h) = \varepsilon h^2$ and for $(t,x) \in Q(h)$ define

$$\mathcal{L}_h u(t,x) = -\frac{u(t,x) - u(t-\varepsilon h^2, x)}{\varepsilon h^2}$$

$$+ \sum_{i=1}^{d} a^{ii}(t,x) \frac{u(t,x+he_i) - 2u(t,x) + u(t,x-he_i)}{h^2}.$$

As above, Assumption 10.5.2 is satisfied due to the same reasons. Assumption 10.5.1 is satisfied obviously. In this method to find $u((k+1)\varepsilon h^2, x)$ from $u(k\varepsilon h^2, x)$, one has to solve an algebraic linear system. On the other hand, there are no restrictions on ε, and one can hope to reach a given time T faster.

10.6. An example

For $d = 1$ we want to find a solution $u(t,x)$ of the problem

$$u_{xx} = u_t \quad \text{in} \quad Q = (0,\infty) \times (0,1), \quad u = g \quad \text{on} \quad \partial' Q, \tag{10.6.1}$$

where g is a given function. We want to find the solution explicitly. In order to be sure that the solution does exist and belongs to $C^{1+\delta/2,2+\delta}(Q)$, we assume that $g \in C^{1+\delta/2,2+\delta}(Q)$ and g vanishes in some neighborhoods of the points $(0,0)$, $(0,1)$.

1. At first assume that $g = 0$ on $\partial_x Q$. In this case the solution can be obtained by reflection. We continue $g(x)$ to all x so that $g(x)$ and $g(1-x)$ are odd functions of x:

$$g(x) = -g(-x), \quad g(x) = -g(2-x). \tag{10.6.2}$$

The extended g can be represented by the following formula:

$$g(x) = \sum_{n=-\infty}^{\infty} \{(gI_{(0,1)})(2n+x) - (gI_{(0,1)})(2n-x)\}.$$

Now we define

$$G_0(t,x)=\frac{1}{\sqrt{4\pi t}}e^{-\frac{1}{4t}|x|^2},\quad u(t,x)=\int_{-\infty}^{\infty}G_0(t,y)g(x-y)\,dy\quad t>0,$$

so that the function $u(t,x)\exp(-t)$ is the function from Theorem 9.1.2. By this theorem u is infinitely differentiable and satisfies the equation $u_{xx}=u_t$ for $t>0$, and $u(t,x)\to g(x)$ as $t\downarrow 0$. Furthermore, from the explicit formula for u it is easy to see that u possesses properties (10.6.2). This implies that $u(t,0)=u(t,1)=0$ for $t>0$. Therefore, we have found a solution of (10.6.1) in our particular case. In addition, we know that this solution is unique. For the future it is useful to rewrite the formula for u in the following way:

$$\begin{aligned}u(t,x)&=\int_{-\infty}^{\infty}G_0(t,x-y)g(y)\,dy\\&=\sum_{n=-\infty}^{\infty}\{\int_{-\infty}^{\infty}G_0(t,x-y)(gI_{(0,1)})(2n+y)\,dy\\&\qquad-\int_{-\infty}^{\infty}G_0(t,x-y)(gI_{(0,1)})(2n-y)\,dy\}\\&=\sum_{n=-\infty}^{\infty}\{\int_{-\infty}^{\infty}G_0(t,2n+x-y)g(y)I_{(0,1)}(y)\,dy\\&\qquad-\int_{-\infty}^{\infty}G_0(t,x+y-2n)g(y)I_{(0,1)}(y)\,dy\}\\&=\int_0^1 G(t,x,y)g(y)\,dy,\end{aligned}\tag{10.6.3}$$

where (summation with respect to n goes from $-\infty$ to ∞)

$$G(t,x,y)=\sum_{n=-\infty}^{\infty}\{G_0(t,x-y-2n)-G_0(t,x+y-2n)\}.$$

If $\varepsilon>0$, then, obviously, for $t\in[\varepsilon,\varepsilon^{-1}]$, $x,y\in[0,1]$ the series converges uniformly along with any its derivative with respect to (t,x,y).

From the maximum principle we know that $0\le u\le 1$ if $0\le g\le 1$. It follows that

$$G\ge 0,\qquad \int_0^1 G(t,x,y)\,dy\le 1.\tag{10.6.4}$$

From the definition of G one can also easily get that

$$G(t,x,y)=G(t,y,x)\quad x,y\in[0,1],\qquad G(t,x,y)=0\quad x\in[0,1],y=0,1.$$

The last equality and (10.6.4) imply that

$$G_y(t,x,y) \geq 0 \quad x \in [0,1], y = 0, \quad G_y(t,x,y) \leq 0 \quad x \in [0,1], y = 1. \tag{10.6.5}$$

Finally, the computations in (10.6.3) are valid for any bounded g which along with (9.1.3) show that if g is a bounded function, which is continuous at a point $x_0 \in (0,1)$, then

$$\int_0^1 G(t,x_0,y)g(y)\,dy \to g(x_0) \tag{10.6.6}$$

as $t \downarrow 0$.

2. Now we want to find a formula for the solution u of (10.6.1) when $g(0,x) \equiv 0$. We claim (cf. Theorem 2.3.1) that for $x \in (0,1)$

$$u(t,x) = \int_0^t [G_y(t-s,x,0)g(s,0) - G_y(t-s,x,1)g(s,1)]\,ds. \tag{10.6.7}$$

To prove this observe that for $s < t$ (Green's formula)

$$G_y(t-s,x,0)u(s,0) - G_y(t-s,x,1)u(s,1)$$

$$= \int_0^1 [G(t-s,x,y)u_{yy}(s,y) - G_{yy}(t-s,x,y)u(s,y)]\,dy.$$

It follows that

$$G_y(t-s,x,0)g(s,0) - G_y(t-s,x,1)g(s,1) = \frac{\partial}{\partial s}\int_0^1 G(t-s,x,y)u(s,y)\,dy. \tag{10.6.8}$$

If g is nonnegative, on the left we have a sum of two positive terms (see (10.6.5)). Therefore (10.6.8) implies that the left-hand side is integrable, and by integrating (10.6.8) we obtain that the right-hand side of (10.6.7) equals

$$\lim_{\varepsilon\downarrow 0}[\int_0^1 G(\varepsilon,x,y)u(t-\varepsilon,y)\,dy - \int_0^1 G(t-\varepsilon,x,y)u(\varepsilon,y)\,dy],$$

which gives (10.6.7) due to (10.6.6).

Thus, we obtain the following representation of the solution to (10.6.1) in the general case of sufficiently regular g vanishing near $(0,0)$, $(0,1)$: for $t > 0, x \in (0,1)$

$$u(t,x) = \int_0^1 G(t,x,y)g(0,y)\,dy$$

$$+ \int_0^t [G_y(t-s,x,0)g(s,0) - G_y(t-s,x,1)g(s,1)]\,ds. \tag{10.6.9}$$

Actually representation (10.6.9) is true for any $g \in C^{1+\delta/2,2+\delta}(Q)$ satisfying the consistency condition: $g_{xx} = g_t$ on $\partial_{tx}Q$. Indeed, in this case we have a solution $u \in C^{1+\delta/2,2+\delta}(Q)$, and we have (10.6.8), which again by integration yields (10.6.9).

Interestingly enough, this representation formula can be used for investigating properties of G. For instance, for $g \equiv 1$ we have $u \equiv 1$; that is, for any $(t,x) \in Q$

$$\int_0^1 G(t,x,y)\,dy + \int_0^t [G_y(t-s,x,0) + |G_y(t-s,x,1)|]\,ds = 1.$$

EXERCISE 10.6.1. From the properties of G stated above it follows that for any g which is integrable over $\partial' Q$ formula (10.6.9) defines a function which is infinitely differentiable in Q and satisfies $u_{xx} = u_t$. Assume that g is bounded and continuous on $(\partial' Q) \setminus \partial_{t,x} Q$ and define u in Q by (10.6.9) and $u = g$ on $\partial' Q$. Prove that u is continuous in $\bar{Q} \setminus \partial_{tx} Q$. Notice that uniqueness of a function which is continuous in $\bar{Q} \setminus \partial_{tx} Q$ and satisfies the heat equation in Q is known from Exercise 8.1.13.

EXERCISE 10.6.2. Take $f \in C^{\delta/2,\delta}(Q)$ and let $f(0,0) = f(0,1) = 0$. Prove that

$$\int_0^t \int_0^1 G(t-s,x,y) f(s,y)\,dyds \tag{10.6.10}$$

defines (or coincides with) a unique solution $u \in C^{1+\delta/2,2+\delta}(Q)$ of the equation $u_{xx} - u_t + f = 0$ with zero boundary condition on $\partial' Q$.

EXERCISE 10.6.3. In the previous exercise drop the condition $f(0,0) = f(0,1) = 0$. By using Exercise 10.4.2 prove that function (10.6.10) is continuous in $\bar{Q}$, equals 0 on $\partial' Q$ and belongs to $C^{1+\delta/2,2+\delta}(Q^\rho)$ for any $\rho > 0$.

10.7. Hints to exercises

10.3.7. Repeat the proof of Theorem 8.11.1.
10.4.3. Calculate $(L - \partial/\partial t)(u - v)$, then apply Chebyshev's inequality (see Exercise 8.1.20).
10.6.1. Remember (10.6.6) and prove a similar fact for the lateral boundary by following the proof of (9.4.5).
10.6.2. Take a bounded and continuous function ϕ on $(0,1)$, fix $t_0 > 0$ and for $t < t_0$ define

$$v(t,x) = \int_0^1 G(t_0 - t, x, y)\phi(y)\,dy.$$

Observe that $v_t + v_{xx} = 0$ and prove that

$$\int_0^1 \phi(y)\Big(\int_0^{t_0}\int_0^1 G(t_0 - t, x, y) f(t,x)\,dxdt\Big)\,dy$$

$$= \int_0^{t_0}\int_0^1 v[u_t - u_{xx}]\,dxdt = \int_0^1 \phi(x) u(t_0, x)\,dx.$$

Bibliography

[1] S. Agmon, A. Douglis and L. Nirenberg, Estimates near the boundary for solutions of elliptic partial differential equations satisfying general boundary conditions, I, Comm. Pure Appl. Math., Vol. 12, 1959, 623–727.

[2] L. Bers, F. John and M. Schechter, "Partial differential equations", Interscience Publishers–John Wiley&Sons, New York–London–Sydney, 1964.

[3] L.A. Caffarelli and X. Cabré, "Fully nonlinear elliptic equations", Colloquium Publications, Vol. 43, American Math. Soc., Providence, RI, 1995.

[4] M. Giaquinta, "Introduction to regularity theory for nonlinear elliptic systems", Birkhauser, Basel–Boston–Berlin, 1993.

[5] F. John, "Partial Differential Equations", 4th ed., Springer-Verlag, New York, 1982.

[6] D. Gilbarg and N.S. Trudinger, "Elliptic partial differential equations of second order", 2d ed., Springer Verlag, Berlin, 1983.

[7] A. Friedman, "Partial differential equations", Holt, Reinhart and Winston, New York, 1969.

[8] N.V. Krylov, "Nonlinear elliptic and parabolic equations of second order", Nauka, Moscow, 1985 in Russian; English translation: Reidel, Dordrecht, 1987.

[9] N.V. Krylov, "Introduction to the theory of diffusion processes", Amer. Math. Soc., Providence, RI, 1995.

[10] O.A. Ladyzhenskaya and N.N. Ural'tseva, "Linear and quasi-linear elliptic equations", Nauka, Moscow, 1964 in Russian; English translation: Academic Press, New York, 1968; 2nd Russian ed. 1973.

[11] O.A. Ladyzhenskaya, V.A. Solonnikov and N.N. Ural'tseva, "Linear and quasi-linear equations of parabolic type", Nauka, Moscow, 1967 in Russian; English translation: American Math. Soc., Providence, RI, 1968.

Index